This report contains the collective views of an international group of experts and does not necessarily represent the decisions or the stated policy of the United Nations Environment Programme, the International Labour Organisation, or the World Health Organization.

Environmental Health Criteria 172

TETRABROMOBISPHENOL A AND DERIVATIVES

First draft prepared by Dr G.J. van Esch,
Bilthoven, Netherlands

Published under the joint sponsorship of
the United Nations Environment Programme,
the International Labour Organisation,
and the World Health Organization

World Health Organization
Geneva, 1995

The **International Programme on Chemical Safety (IPCS)** is a joint venture of the United Nations Environment Programme, the International Labour Organisation, and the World Health Organization. The main objective of the IPCS is to carry out and disseminate evaluations of the effects of chemicals on human health and the quality of the environment. Supporting activities include the development of epidemiological, experimental laboratory, and risk-assessment methods that could produce internationally comparable results, and the development of manpower in the field of toxicology. Other activities carried out by the IPCS include the development of know-how for coping with chemical accidents, coordination of laboratory testing and epidemiological studies, and promotion of research on the mechanisms of the biological action of chemicals.

WHO Library Cataloguing in Publication Data

Tetrabromobisphenol A and derivatives.

(Environmental health criteria ; 172)

1.Bromine compounds 2.Environmental exposure
3.Occupational exposure 4.Flame retardants I.Series

ISBN 92 4 157172 1 (NLM Classification: QD 181.B7)
ISSN 0250-863X

PRINTED IN FINLAND
95/10610 – VAMMALA – 5500

CONTENTS

ENVIRONMENTAL HEALTH CRITERIA FOR TETRABROMOBISPHENOL A AND DERIVATIVES

TETRABROMOBISPHENOL A DERIVATIVES

NOTE TO READERS OF THE CRITERIA MONOGRAPHS

Every effort has been made to present information in the criteria monographs as accurately as possible without unduly delaying their publication. In the interest of all users of the Environmental Health Criteria monographs, readers are requested to communicate any errors that may have occurred to the Director of the International Programme on Chemical Safety, World Health Organization, Geneva, Switzerland, in order that they may be included in corrigenda.

* * *

A detailed data profile and a legal file can be obtained from the International Register of Potentially Toxic Chemicals, Case postale 356, 1219 Châtelaine, Geneva, Switzerland (Telephone No. 9799111).

* * *

This publication was made possible by grant number 5 U01 ES02617-15 from the National Institute of Environmental Health Sciences, National Institutes of Health, USA, and by financial support from the European Commission.

Environmental Health Criteria

P R E A M B L E

Objectives

In 1973 the WHO Environmental Health Criteria Programme was initiated with the following objectives:

(i) to assess information on the relationship between exposure to environmental pollutants and human health, and to provide guidelines for setting exposure limits;

(ii) to identify new or potential pollutants;

(iii) to identify gaps in knowledge concerning the health effects of pollutants;

(iv) to promote the harmonization of toxicological and epidemiological methods in order to have internationally comparable results.

The first Environmental Health Criteria (EHC) monograph, on mercury, was published in 1976 and since that time an ever-increasing number of assessments of chemicals and of physical effects have been produced. In addition, many EHC monographs have been devoted to evaluating toxicological methodology, e.g., for genetic, neurotoxic, teratogenic and nephrotoxic effects. Other publications have been concerned with epidemiological guidelines, evaluation of short-term tests for carcinogens, biomarkers, effects on the elderly and so forth.

Since its inauguration the EHC Programme has widened its scope, and the importance of environmental effects, in addition to health effects, has been increasingly emphasized in the total evaluation of chemicals.

The original impetus for the Programme came from World Health Assembly resolutions and the recommendations of the 1972 UN Conference on the Human Environment. Subsequently the work became an integral part of the International Programme on Chemical Safety (IPCS), a cooperative programme of UNEP, ILO and WHO. In this manner, with the strong support of the new partners, the importance of occupational health and environmental effects was fully recognized. The EHC monographs have become widely established, used and recognized throughout the world.

The recommendations of the 1992 UN Conference on Environment and Development and the subsequent establishment of the Intergovernmental Forum on Chemical Safety with the priorities for action in the six programme areas of Chapter 19, Agenda 21, all lend further weight to the need for EHC assessments of the risks of chemicals.

Scope

The criteria monographs are intended to provide critical reviews on the effect on human health and the environment of chemicals and of combinations of chemicals and physical and biological agents. As such, they include and review studies that are of direct relevance for the evaluation. However, they do not describe *every* study carried out. Worldwide data are used and are quoted from original studies, not from abstracts or reviews. Both published and unpublished reports are considered and it is incumbent on the authors to assess all the articles cited in the references. Preference is always given to published data. Unpublished data are only used when relevant published data are absent or when they are pivotal to the risk assessment. A detailed policy statement is available that describes the procedures used for unpublished proprietary data so that this information can be used in the evaluation without compromising its confidential nature (WHO (1990) Revised Guidelines for the Preparation of Environmental Health Criteria Monographs. PCS/90.69, Geneva, World Health Organization).

In the evaluation of human health risks, sound human data, whenever available, are preferred to animal data. Animal and *in vitro* studies provide support and are used mainly to supply evidence missing from human studies. It is mandatory that research on human subjects is conducted in full accord with ethical principles, including the provisions of the Helsinki Declaration.

The EHC monographs are intended to assist national and international authorities in making risk assessments and subsequent risk management decisions. They represent a thorough evaluation of risks and are not, in any sense, recommendations for regulation or standard setting. These latter are the exclusive purview of national and regional governments.

Content

The layout of EHC monographs for chemicals is outlined below.

- Summary - a review of the salient facts and the risk evaluation of the chemical
- Identity - physical and chemical properties, analytical methods
- Sources of exposure
- Environmental transport, distribution and transformation
- Environmental levels and human exposure
- Kinetics and metabolism in laboratory animals and humans
- Effects on laboratory mammals and *in vitro* test systems
- Effects on humans
- Effects on other organisms in the laboratory and field
- Evaluation of human health risks and effects on the environment
- Conclusions and recommendations for protection of human health and the environment
- Further research
- Previous evaluations by international bodies, e.g., IARC, JECFA, JMPR

Selection of chemicals

Since the inception of the EHC Programme, the IPCS has organized meetings of scientists to establish lists of priority chemicals for subsequent evaluation. Such meetings have been held in: Ispra, Italy, 1980; Oxford, United Kingdom, 1984; Berlin, Germany, 1987; and North Carolina, USA, 1995. The selection of chemicals has been based on the following criteria: the existence of scientific evidence that the substance presents a hazard to human health and/or the environment; the possible use, persistence, accumulation or degradation of the substance shows that there may be significant human or environmental exposure; the size and nature of populations at risk (both human and other species) and risks for environment; international concern, i.e. the substance is of major interest to several countries; adequate data on the hazards are available.

If an EHC monograph is proposed for a chemical not on the priority list, the IPCS Secretariat consults with the Cooperating Organizations and all the Participating Institutions before embarking on the preparation of the monograph.

Procedures

The order of procedures that result in the publication of an EHC monograph is shown in the flow chart. A designated staff member of IPCS, responsible for the scientific quality of the document, serves as Responsible Officer (RO). The IPCS Editor is responsible for layout and language. The first draft, prepared by consultants or, more usually, staff from an IPCS Participating Institution, is based initially on data provided from the International Register of Potentially Toxic Chemicals, and reference data bases such as Medline and Toxline.

The draft document, when received by the RO, may require an initial review by a small panel of experts to determine its scientific quality and objectivity. Once the RO finds the document acceptable as a first draft, it is distributed, in its unedited form, to well over 150 EHC contact points throughout the world who are asked to comment on its completeness and accuracy and, where necessary, provide additional material. The contact points, usually designated by governments, may be Participating Institutions, IPCS Focal Points, or individual scientists known for their particular expertise. Generally some four months are allowed before the comments are considered by the RO and author(s). A second draft incorporating comments received and approved by the Director, IPCS, is then distributed to Task Group members, who carry out the peer review, at least six weeks before their meeting.

The Task Group members serve as individual scientists, not as representatives of any organization, government or industry. Their function is to evaluate the accuracy, significance and relevance of the information in the document and to assess the health and environmental risks from exposure to the chemical. A summary and recommendations for further research and improved safety aspects are also required. The composition of the Task Group is dictated by the range of expertise required for the subject of the meeting and by the need for a balanced geographical distribution.

The three cooperating organizations of the IPCS recognize the important role played by nongovernmental organizations. Representatives from relevant national and international associations may be invited to join the Task Group as observers. While observers may provide a valuable contribution to the process, they can only speak at the invitation of the Chairperson.

EHC PREPARATION FLOW CHART

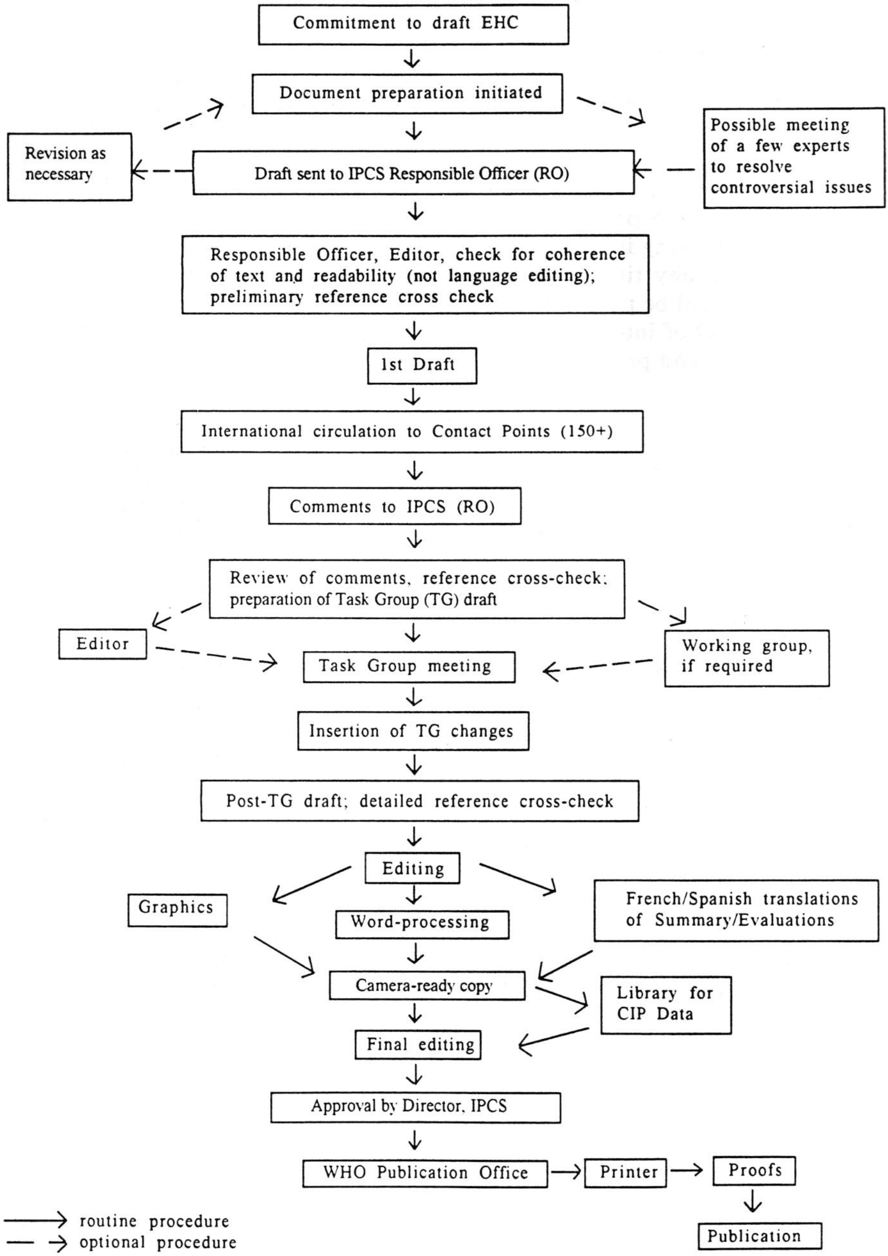

Observers do not participate in the final evaluation of the chemical; this is the sole responsibility of the Task Group members. When the Task Group considers it to be appropriate, it may meet *in camera*.

All individuals who as authors, consultants or advisers participate in the preparation of the EHC monograph must, in addition to serving in their personal capacity as scientists, inform the RO if at any time a conflict of interest, whether actual or potential, could be perceived in their work. They are required to sign a conflict of interest statement. Such a procedure ensures the transparency and probity of the process.

When the Task Group has completed its review and the RO is satisfied as to the scientific correctness and completeness of the document, it then goes for language editing, reference checking, and preparation of camera-ready copy. After approval by the Director, IPCS, the monograph is submitted to the WHO Office of Publications for printing. At this time a copy of the final draft is sent to the Chairperson and Rapporteur of the Task Group to check for any errors.

It is accepted that the following criteria should initiate the updating of an EHC monograph: new data are available that would substantially change the evaluation; there is public concern for health or environmental effects of the agent because of greater exposure; an appreciable time period has elapsed since the last evaluation.

All Participating Institutions are informed, through the EHC progress report, of the authors and institutions proposed for the drafting of the documents. A comprehensive file of all comments received on drafts of each EHC monograph is maintained and is available on request. The Chairpersons of Task Groups are briefed before each meeting on their role and responsibility in ensuring that these rules are followed.

WHO TASK GROUP ON ENVIRONMENTAL HEALTH CRITERIA FOR TETRABROMOBISPHENOL A AND DERIVATIVES

Members

Dr D. Anderson, BIBRA Toxicology International, Carshalton, United Kingdom

Dr R. Benson, Drinking Water Branch, US EPA, Denver, USA

Dr B. Jansson, Institute of Applied Environmental Research, Stockholm University, Solna, Sweden

Dr J. Kielhorn, Fraunhofer Institute for Toxicology and Aerosol Research, Hanover, Germany

Dr R.D. Kimbrough, Institute for Evaluating Health Risks, Washington DC, USA (*Vice-chairman*)

Dr D. Osborn, Institute of Terrestrial Ecology, Monks Wood, Huntingdon, United Kingdom

Dr Wai-On Phoon, Department of Occupational Health, University of Sydney, Sydney, Australia (*Chairman*)

Dr J. Sekizawa, National Institute of Health Sciences, Tokyo, Japan (*Rapporteur*)

Dr E. Söderlund, National Institute of Public Health, Oslo, Norway

Observers

Dr M.L. Hardy, Albemarle Corporation, Baton Rouge, USA

Dr D.L. McAllister, Quality, Environment, Health and Safety, and Research Support, Great Lakes Chemical Corporation, West Lafayette, USA

Secretariat

Dr K.W. Jager, International Programme on Chemical Safety, World Health Organization, Geneva, Switzerland (*Secretary*)

ENVIRONMENTAL HEALTH CRITERIA FOR TETRABROMOBISPHENOL A AND DERIVATIVES

A WHO Task Group on Environmental Health Criteria for Tetrabromobisphenol A and Derivatives met at BIBRA Toxicology International, Carshalton, United Kingdom, from 6 to 11 June 1994. Dr K.W. Jager, IPCS, welcomed the participants on behalf of Dr M. Mercier, Director of the IPCS, and the three IPCS cooperating organizations (UNEP/ILO/WHO). The Group reviewed and revised the draft and made an evaluation of the risks for human health and the environment from exposure to Tetrabromobisphenol A and derivatives.

The first draft was prepared by Dr G.J. van Esch, the Netherlands, who also prepared the second draft, incorporating comments received following circulation of the first drafts to the IPCS Contact Points for Environmental Health Criteria monographs.

Dr K.W. Jager of the IPCS Central Unit was responsible for the scientific content of the monograph and Mrs M.O. Head of Oxford for the technical editing.

The fact that industry made available to the IPCS and the Task Group their proprietary toxicological information on their products under discussion is gratefully acknowledged. This allowed the members of Task Group to make their evaluation on a more complete data base.

The efforts of all who helped in the preparation and finalization of the monograph are gratefully acknowledged.

INTRODUCTION

Tetrabromobisphenol A (TBBPA) is an important flame retardant. The demand for tetrabromobisphenol A and its derivatives accounts for over 60 000 tonnes per year.

Whatever their use, flame retardants will ultimately end up in the environment, as such, or as break-down products. In the case of tetrabromobiphenol A, the ultimate breakdown products and their levels may be different, depending on whether TBBPA has been used as a reactive or as an additive flame retardant.

In order to make a proper assessment of the hazards of a substance for humans and the environment, it is essential that data are available not only on toxicity and ecotoxicity but also on:

- the ultimate fate of the substance under various use and disposal conditions, including incineration, and on its breakdown products; and

- the persistence and bioaccumulation/biomagnification of the substance and its breakdown products.

The IPCS is preparing several EHC monographs on Flame Retardants, which will provide additional information relevant to TBBPA.

One monograph, Flame retardants - General introduction (in preparation), will include a general introduction to the uses, the modes of action, and the potential risks of flame retardants, and also a list of the substances used as flame retardants with a general indication of the data available on them.

Flame retardants in wide use are discussed in separate monographs, e.g., EHC 162: Polybrominated Diphenyl Ethers.

Some flame retardants, considered hazardous for humans and the environment, have also been reviewed in separate monographs including EHC 152: Polybrominated Biphenyls, and EHC 173: Tris- and bis(2,3-dibromopropyl) Phosphate.

Because of the possibility of the formation of halogenated dibenzodioxins and dibenzofurans under certain circumstances, such as pyrolysis, the following monographs have been developed:

EHC 88: Polychlorinated dibenzo-*para*-dioxins and dibenzofurans and Polybrominated dibenzodioxins and dibenzofurans (in preparation).

The reader should consult these monographs for further information.

TETRABROMOBISPHENOL A

1. SUMMARY AND EVALUATION; CONCLUSIONS AND RECOMMENDATIONS ON TETRABROMOBISPHENOL A (TBBPA)

1.1 Summary and evaluation

1.1.1 Physical and chemical properties

TBBPA is a white (colourless), crystalline powder, containing 59% bromine. The melting point is approximately 180 °C and the boiling point, 316 °C. Vapour pressure is much less than 1 mmHg at 20 °C. TBBPA has a low solubility in water, but is very soluble in methanol and acetone. The *n*-octanol/water partition coefficient (log P_{ow}) is 4.5.

1.1.2 Production and use

Commercial TBBPA is the brominated flame retardant produced in the largest amounts globally. The demand for TBBPA and its derivatives accounts for over 60 000 tonnes per year. TBBPA is used as a reactive (primary use) or additive flame retardant in polymers, such as ABS, epoxy and polycarbonate resins, high impact polystyrene, phenolic resins, adhesives, and others.

1.1.3 Environmental transport, distribution, and transformation

Because of its partition coefficient and low water solubility, TBBPA in the environment is expected to sorb to a large extent onto sediment and organic matter in the soil.

Accumulation studies on aquatic invertebrates and vertebrates indicate bioconcentration factors ranging from 20 to 3200. The half-life in fish is less than 1 day, and that in oysters, less than 5 days. During depuration, most of the accumulated TBBPA (and metabolites) will be eliminated within 3-7 days.

Biodegradation studies showed that TBBPA is partly degraded under both aerobic and anaerobic conditions, in soil, and in river sediment and water. Depending on soil type, temperature, humidity, and the composition of the soil, approximately 40-90% of TBBPA remained in the soils after 56-64 days. Under sewage treatment conditions, no biodegradation was detected, as measured as BOD, in 2 weeks.

Laboratory pyrolysis studies showed that polymers with TBBPA, with and without the presence of Sb_2O_3, at different temperatures, in the presence of oxygen, etc., may form polybrominated dibenzofurans (PBDF) and, to a lesser extent, polybrominated dibenzodioxins (PBDD). Mainly lower brominated PBDF and PBDD are formed. When polymers formulated with TBBPA, exposed to simulating thermal processing conditions, were analysed, 2,3,7,8-PBDD/PBDF were not detected. Only mono- or dibromo-substituted PBDF were detected at up to 100 μg/kg levels in the resin. Investigation of the workplace atmosphere showed no 2,3,7,8-substituted PBDD/PBDF (detection limit = 0.1 ng/m^3).

In recycled TBBPA-containing polymers, less than 5 μg total PBDF/PBDD per kg were detected and 2,3,7,8-substituted congeners were only found at levels of less than 0.2 μg/kg.

In a warehouse fire, in which a great quantity of polybutylene terephthalate (PBT) containing TBBPA was burnt, only low levels of 2,3,7,8-substituted tetra-, penta-, and hexa-BDF/BDD (less than 5 μg/kg) were detected in burnt PBT and ash/slag samples.

1.1.4 *Environmental levels and human exposure*

TBBPA was detected in some sediments in Japan and Sweden and in fish (2 samples near an industrialized area out of 229 samples) in μg/kg levels in Japan. The dimethoxy derivative of TBBPA could be identified in mussels and sediment. TBBPA was not generally detected in water.

1.1.5 *Kinetics and metabolism in laboratory animals and humans*

In rats, TBBPA is poorly absorbed from the gastrointestinal tract. Once absorbed, it and/or its metabolites appear to be distributed throughout most organs of the body. In the rat, the maximum half-life in any tissue was less than 2 1/2 days.

1.1.6 *Effects on laboratory mammals and* in vitro *test systems*

The acute oral toxicity of TBBPA for laboratory animals is low. The oral LD_{50} for the rat was > 5 g/kg body weight and the oral LD_{50} for the mouse was 10 g/kg body weight. The dermal LD_{50} for the rabbit was > 2 g/kg body weight. The inhalation LC_{50}s for the mouse, rat, and guinea-pig were > 0.5 mg/litre. A single dermal application of TBBPA on the skin of rabbits and guinea-pigs did not induce local or systemic effects at concentrations of up to 3.16 g/kg body weight. TBBPA was not irritating to rabbit skin or eyes. No sensitization reaction was observed in a few studies on guinea-pigs. TBBPA was also tested for chloracnegenic activity in rabbit ears. No such reaction was observed. A 3-week dermal toxicity study, in which the clipped and abraded skin of rabbits was exposed to up to 2500 mg TBBPA/kg body weight, showed only slight skin erythema. No other compound-related changes were observed.

Rats were exposed to up to 18 mg micronized TBBPA/litre (18 000 mg/m^3) for 4 h/day, 5 days/week for 2 weeks. No effects on body weight, histopathology, haematology, serum chemistry, or urinalysis were observed.

Oral doses to rats of up to 1000 mg TBBPA/kg diet for 28 days did not produce any adverse effects. The total bromine contents of the liver did not differ between the control and high-dose (1000 mg/kg) groups.

In an oral, 90-day toxicity study on rats, dose levels of up to 100 mg TBBPA/kg body weight did not induce any adverse effects on body weight, haematology, clinical chemistry, urinalysis, organ weights, or gross and microscopic examinations.

In an oral, 90-day study on mice, a dose of 4900 mg/kg diet (approximately 700 mg/kg body weight per day) did not cause any adverse effects; a dose of 15 600 mg/kg diet (approximately 2200 mg/kg body weight per day) caused decreased body weight, increased spleen weight, and reduced concentration of red blood cells, serum proteins, and serum triglyceride.

Two teratogenicity studies were carried out on rats; one in which dose levels of up to 10 g/kg body weight were administered by gavage from gestation day 6 to day 15 and a second in which dose levels of up to 2.5 g/kg body weight were administered from day 0 to day 19 of gestation. In the first study, 3/5 animals receiving 10 g/kg died, but no signs of toxicity were noticed in animals receiving 3 g/kg. No teratogenic effects were observed. No abnormalities were found in the second study.

TBBPA was not mutagenic in various studies with *Salmonella typhimurium* strains TA1535, TA1537, TA1538, TA98, and TA100 with metabolic activation by an S9 mix of Aroclor-induced rats and Syrian hamsters. The concentrations tested were up to 10 000 μg/plate. The results of two tests with *Saccharomyces cerevisiae*, with and without microsomal enzyme preparation from Aroclor-induced rats, were also negative.

No carcinogenicity or long-term toxicity studies were reported.

1.1.7 Effects on humans

TBBPA did not produce any skin irritation or sensitization in 54 human volunteers.

No epidemiological studies or other data on the effects on humans are available.

1.1.8 Effects on other organisms in the laboratory and field

TBBPA was not very toxic for marine algae. In 28 short-term studies, the EC_{50}s were in the range of 0.1-1.0 mg/litre, while fresh water algae did not show growth inhibition, even at 9.6 mg/litre.

An acute 48-h LC_{50} for *Daphnia magna* was reported to be 0.96 mg/litre; at 0.32 mg/litre, 5% of the organisms died. In a 21-day study, however, the EC_{50} for survival and growth of *Daphnia magna* was > 0.98 mg/litre. Based on the effects of TBBPA on daphnid reproduction in this 21-day study, a Maximum Acceptable Toxicant Concentration (MATC) was between 0.30 and 0.98 mg/litre. Mysid shrimp (age < 1, 5, and 10 days old) showed 96-h LC_{50} values of 0.86, 1.1, and 1.2 mg/litre, respectively.

The 96-h acute EC_{50} (reduction of shell deposition) in Eastern oysters was calculated to be 0.098 mg/litre with a no-observed-effect concentration (NOEC) of 0.0062 mg/litre.

The 96-h LC_{50}s of TBBPA for bluegill sunfish, rainbow trout, and fathead minnow were 0.51, 0.40, and 0.54 mg/litre, respectively. The no-effect concentrations for these three fish species were 0.10, 0.18, and 0.26 mg/litre. Fathead minnow (embryos and larvae) were exposed for 35 days to TBBPA and showed a MATC of between 0.16 and 0.31 mg/litre, based on adverse effects on embryo and larvae survival.

The 14-day, no-effect levels for the sediment invertebrate midge *Chironomous tentans* were 0.039, 0.045, and 0.046 mg TBBPA/litre water in low, medium, and high organic carbon sediments, respectively.

Most of the studies on aquatic systems have been performed at pHs around the pKa_2. The behaviour of TBBPA in acidic waters may be different.

1.2 Conclusions

1.2.1 General population

TBBPA is widely used and incorporated in polymers as a reactive or additive flame retardant. Contact of the general population is with products made from these polymers and would not result in significant uptake of TBBPA. Furthermore, the acute and repeated dose toxicity of TBBPA is very low. TBBPA is poorly absorbed from the gastrointestinal tract. The risk for the general population from TBBPA exposure is, therefore, considered to be insignificant.

1.2.2 Occupational exposure

Occupational exposure to TBBPA is primarily as particulates during packaging or mixing operations. The control of dust through the use of local ventilation and other engineering methods will reduce the risk to workers. If dust cannot be adequately controlled, respiratory protection should be used.

1.2.3 The environment

Where detected in the environment, TBBPA is mainly found in soil and sediment samples. A relatively high bioconcentration factor seems to be balanced by rapid excretion and the compound has not normally been found in environmental biological samples.

The phenolic groups of TBBPA may be methylated in the environment and the resulting Me_2-TBBPA is more lipophilic. This compound has also been found in sediment, fish, and shellfish.

1.2.4 Breakdown products

PBDD and PBDF have been found as trace impurities in TBBPA; however, the presence of 2,3,7,8-congeners has not been demonstrated. Under laboratory pyrolysis conditions, PBDF/PBDD are formed from TBBPA.

A limited number of studies have shown that only trace quantities of PBDF/PBDD may be produced during the processing and recycling of polymers containing TBBPA as an additive flame retardant. Proper ventilation and other engineering controls can prevent worker exposure.

1.3 Recommendations

1.3.1 General

- Workers in the manufacture of TBBPA and products containing the compound should be protected from exposure by means of engineering controls, monitoring of occupational exposure, and appropriate industrial hygiene measures.

- Environmental exposure should be minimized through the appropriate treatment of effluents and emissions in industries using the compound or products.

- Disposal of industrial wastes and consumer products should be controlled to minimize environmental contamination with this material and its breakdown products.

- If TBBPA-treated material is incinerated, it has to be done in properly constituted incinerators running at consistently optimal conditions.

1.3.2 Further studies

- Monitoring of environmental samples for TBBPA, Me_2-TBBPA, and PBDF/PBDD should be continued, and if these compounds are found, human monitoring should also be carried out.

- Monitoring should be conducted to measure occupational exposures to respirable particles of TBBPA; if indicated by workplace monitoring, a short-term inhalation study on rats should be conducted.

- Studies on PBDF/PBDD formation from TBBPA-treated material during incineration, accidental fires, and under conditions simulating fire, should be conducted.

- Long-term studies of the fate of polymers containing TBBPA (both added and reacted into the polymer), especially in land fills, should be conducted.

- Environmental conversion of TBBPA to its dimethyl derivative, especially in sediments, should be studied.

- Studies on the recyclability of TBBPA-containing polymers should be continued, paying attention to break-down products.

- Since there are no data, an additional *in vitro* test with TBBPA for cytogenetic damage is required. If this test is positive, additional *in vivo* studies will be necessary. If the cytogenetic testing *in vivo* shows positive results, additional short- or long-term testing is required.

- Since there are no data, a test for reproductive toxicity in rats is required.

2. IDENTITY, PHYSICAL AND CHEMICAL PROPERTIES, ANALYTICAL METHODS

2.1 Identity

Chemical formula	$C_{15}H_{12}Br_4O_2$
Chemical structure	

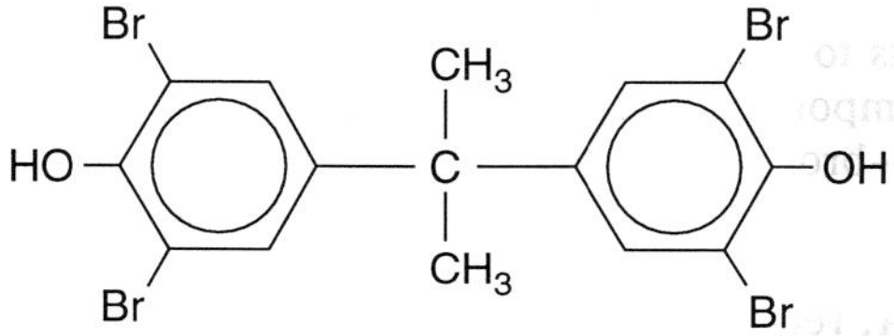

Relative molecular mass	543.92
Chemical name	phenol, 4,4′-(1-methylethylidene) *bis*[2,6-dibromo-]
Common abbreviation	TBBPA
CAS registry number	79-94-7
EINECS number	2012369
Synonyms	4,4′-isopropylidene-bis(2,6-dibromophenol); 2,2-bis(3,5-dibromo-4-hydroxyphenyl) propane; phenol, 4,4′-isopropylidenebis (dibromo-); 3,3′,5,5′-tetrabromobisphenol A; tetrabromodian: tetrabromodihydroxy diphenylpropane.

2.1.1 *Technical product*

Trade names	Great Lakes BA-59P; Saytex RB-100; Saytex RB-100 ABS; FR-1524; Bromdian; FG 2000; Fire Guard 2000; Firemaster BP 4A; Tetrabrom

2.2 Physical and chemical properties

Tetrabromobisphenol A (TBBPA) is a white (colourless), crystalline or powdered solid with a slight characteristic odour, containing 58.7% bromine.

The purity of commercial TBBPA is 98.5% containing 0.1% water, a maximum of 60 mg hydrolysable bromine/kg, and a maximum of 100 mg ionic bromide/kg (Ethyl Corporation, 1992b).

In experiments to determine levels of breakdown products in the technical compounds, Thoma et al. (1986a) found hexa-, penta-, and octa-brominated dibenzofurans (12, 31, 19 µg/kg, respectively) in a sample of technical grade TBBPA. Thies et al. (1990), measuring mono- to hexa-BDF/BDD in a commercial sample of TBBPA, reported a total of less than 20 µg/kg; neither 2,3,7,8-TeBDD nor 2,3,7,8-TeBDF was detectable (detection limit, 0.5 µg/kg).

In an ultratrace analysis specific for 15 different PBDF/PBDD having bromine in the 2,3,7,8-positions (Tondeur et al., 1990), none of these specific congeners were found in TBBPA (limit of determination for the 15 different 2,3,7,8-substituted PBDF/PBDD ranged from 0.1 up to 1000 µg/kg) (Ranken, 1993; Remmers et al., 1993).

Physical and chemical properties of commercial products are summarized in Table 1.

2.3 Conversion factor for air concentrations

1 ppm = 0.02 mg/litre under standard conditions (Bayer, 1990).

2.4 Analytical methods

TBBPA can be converted to the diethyl derivative by ethylation, and the resulting product can be determined by gas chromatographic (GC) and gas chromatographic/mass spectrometric (GC/MS) analysis (Gustafsson & Wallen, 1988).

Seawater samples were acidified with HCl, extracted with petroleum ether, concentrated, and taken up in hexane. Quantification was performed with on-column capillary gas chromatography using an electron-capture detector. The response of the detector was linear from 0.50 to 5.0 ng. The limit of determination was 1.0 μg/litre. Recovery of fortified TBBPA at 535 μg/litre was 99%, and, at 84 μg/litre, 93 ± 11.5% (Goodman et al., 1988).

Table 1. Physical and chemical properties of commercial products

Melting point	181-182 °C[a,b]
Boiling point	316 °C (approximately)[b]
Specific gravity	2.18[a]
Flash point	178 °C[b]
Vapour pressure	< 1 mmHg at 20 °C[b]
Solubility in water	0.72 mg/litre at 15 °C* (< 0.1 wt% at 25 °C)[b] 4.16 mg/litre at 25 °C*[b] 1.77 mg/litre at 35 °C*[b] in methanol, 920 g/litre (47.2 wt% at 25 °C)[a] in acetone, 2400 g/litre (69.6 wt% at 25 °C)[a] in toluene 6.4 wt% at 25 °C[a] in styrene < 1.0 wt% at 25 °C[a]
n-Octanol/water partition coefficient (log P_{ow})	4.5-5.3[b]
pKa_1 and pKa_2	7.5 and 8.5, respectively[b]

[a] From: Ethyl Corporation (1992a,b); Great Lakes Chemical Corporation (1986).
[b] From: Bayer (1990).
* The water solubility was determined by radioassay, using (phenyl-UL-^{14}C) labelled TBBPA.

TBBPA determination in freshwater was carried out by HPLC analysis. A C_{18} column was used and the mobile phase was a 80/20 acetonitrile/HPLC grade water mixture using a UV detector (wavelength 230 nm). The average recovery of TBBPA from water was 96.1%. The theoretical, minimum detectable concentration was < 2.22 μg/ml active ingredient (Surprenant, 1988, 1989a).

Watanabe et al. (1983a) described an analytical method to determine TBBPA in sediment. The TBBPA in the extract was converted to a diethylether derivative by ethylation. This derivative was identified and determined by gas chromatography (ECD-^{63}Ni detector) and gas chromatography/mass spectrometry. This extraction, clean-up, and determination method was also used to determine TBBPA in mussel tissue. A method was also given to determine methoxy-TBBPA in mussel tissue.

3. SOURCES OF HUMAN AND ENVIRONMENTAL EXPOSURE

3.1 Natural occurrence

TBBPA is not reported to occur naturally.

3.2 Anthropogenic sources

3.2.1 *Production levels and processes*

TBBPA is the largest selling brominated flame retardant accounting for an annual market of 41 000 tonnes (Japan - 15 000 tonnes, USA - 16 000 tonnes, and Europe 10 000 tonnes) (OECD, 1993). About 10 000 tonnes/year of TBBPA derivatives are produced (Satoh & Sugie, 1993). In the USA, 5000-6350 tonnes were reported to have been used in 1982.

In 1987, Sweden imported more than 100 tonnes, and the Netherlands consumed 200 tonnes in 1988. The USA imported 660 tonnes in 1983 and produced 39 000 tonnes in 1983/1984 (Gustafsson & Wallen, 1988). The annual consumption of TBBPA in Japan was 14 400 tonnes in 1987, 18 000 tonnes in 1988, 23 000 tonnes in 1990, 24 500 tonnes in 1991, 23 000 tonnes in 1992, and 22 000 tonnes in 1993, mainly for use in flame retarding polystyrene (ABS, HIPS) and PC, and, partly for use in the manufacturing of derivatives (Tatsukawa & Watanabe, 1990; Watanabe & Tatsukawa, 1990; The Chemical Daily, 1990-1994).

Tetrabromobisphenol A (TBBPA) is produced by the bromination of bisphenol A (BPA) in the presence of a solvent. The bromination reaction is generally conducted in solvents, such as a halocarbon alone, or with water or 50% hydrobromic acid, or aqueous alkyl monoethers. Aqueous acetic acid is also a satisfactory medium and acetic acid with added sodium acetate is reported to improve product colour (Ullmann, 1985). If methanol is used as a solvent, the fumigant methyl bromide is produced as a co-product. The BPA dissolved in methanol is reacted with bromine to yield TBBPA and hydrobromic acid. The methanol reacts further with the hydrobromic acid to yield methyl bromide. This process is used by Ethyl Corporation and Great Lakes Chemical Corporation (Ethyl Corporation, personal communication, 1990). Dead Sea Bromine, who produce TBBPA at plants in Israel and the Netherlands, do not use this method but another process (Dead Sea Bromine, personal communication, 1994).

Following the synthesis, the TBBPA is precipitated from the methanol, separated by filtration, and washed to remove impurities. The solid product is dried and then packaged in bags, drums, or bulk containers.

The process is largely conducted in enclosed equipment, therefore limiting the possibility of worker exposure. However, some exposure to dust may occur during the packaging process.

3.2.2 Uses

The primary use of TBBPA is as a reactive intermediate in the manufacture of flame-retarded epoxy and polycarbonate resins, accounting for approximately 90% of TBBPA used. The identity of TBBPA is lost in the process of polymerization (McAllister, personal communication, 1994). Polymerization is typically conducted in totally enclosed equipment, minimizing the possibility of worker exposure. A principal use of TBBPA epoxy resins is in printed circuit boards where the bromine content may be 20% by weight (34% TBBPA).

As an additive flame retardant, TBBPA, in the form of a dry powder, is mixed with various polymers. Dusting may occur during mixing. It does not react chemically with the other compounds, and, therefore, may leach out of the polymer matrix. Additive use accounts for approximately 10% of TBBPA used (McAllister, personal communication, 1994). For example, TBBPA may be used as an additive flame retardant in acrylonitrile-butadiene-styrene (ABS) resins and in high impact polystyrene. TBBPA can be used as an additive flame retardant in ABS thermoplastics, in polystyrene, and in phenolic resins. Recommended starting levels of TBBPA in ABS (medium to high impact) are 17.6-22.0% and 14% in high impact polystyrene. ABS resins are used in automotive parts, pipes and fittings, refrigerators, other appliances, business machines, and telephones. Polystyrene is used in packaging, consumer products, disposables, electrical and electronic equipment, furniture, and in building and construction materials (Quast et al., 1975; Personal communication on opportunities for cooperation on acrolein and tetrabromobisphenol A from Gustafsson K & Wallen M of National Chemicals Inspectorate, Scientific Documentation and Research, Solna, Sweden, in 1988).

When used as a flame retardant in encapsulated integrated circuit devices, TBBPA may be combined with an additional flame retardant, such as antimony trioxide. TBBPA can be used as a reactive flame retardant in polycarbonate and unsaturated polyester resins. Polycarbonates are used in communication and electronics equipment (i.e., business machines), appliances, transportation devices, sports and recreation equipment, lighting fixtures and signs.

Unsaturated polyesters are used for making simulated marble floor tiles, bowling balls, glass-reinforced panels, furniture parts, sewer-pipes coupling compound, automotive patching compounds, buttons, and for encapsulating electrical devices (Gustafsson & Wallen, 1988).

TBBPA may also be used as an intermediate for the production of other flame retardants, such as the bis(2-hydroxyethyl ether) of TBBPA, as a flame retardant for paper and textiles, in adhesives and coatings, and for imparting corrosion resistance to unsaturated polyesters used in chemical processing equipment (Gustafsson & Wallen, 1988).

The total use of TBBPA derivatives is only about 25% as large as the use of TBBPA itself (McAllister, personal communication, 1994).

4. ENVIRONMENTAL TRANSPORT, DISTRIBUTION, AND TRANSFORMATION

4.1 Transport and distribution between media

Because of its partition coefficient and low water solubility, TBBPA in the environment is expected to sorb onto sediment and organic matter in soil.

4.2 Transformation

4.2.1 Biotransformation

The dimethyl ether derivative of TBBPA, thought to be a metabolite from microbial methylation, was found in river sediment collected in Osaka, Japan. None was found in marine sediment collected in Osaka Bay, Japan. The dimethylated-TBBPA derivative was detected at about one-hundredth of the TBBPA levels concurrently measured in river sediment (Watanabe et al., 1983b).

Sediment samples taken upstream and downstream from a factory in Sweden were analysed for TBBPA and dimethylated-TBBPA. Up and downstream from the factory, the TBBPA levels were 50 and 430 ng/g ignition loss and those for methylated-TBBPA were 36 and 2400 ng/g ignition loss (Sellström, 1990).

4.2.2 Biodegradation

The biodegradability of ^{14}C-TBBPA was tested under aerobic conditions in three soil types, i.e., Massachusetts sandy loam, a California loam, and Arkansas silty loam. The three soil types contained: sand (83%) silt (13%) clay (4%), sand (16%) silt (58%) clay (26%), and sand (43%) silt (24%) clay (33%), respectively. Thin layer chromatography (TLC) showed biodegradation of TBBPA in all soil types. Only 6% or less of the applied radioactive TBBPA was recovered in the volatile traps, indicating only partial degradation. Results of the TLC analysis indicated variable degradation rates of TBBPA. After 64 days, the amount of TBBPA remaining in the soils ranged from 82 to 36%, with the highest levels in the sandy loam soil and the lowest in the silty loam soil (Fackler, 1989a).

The biodegradability of TBBPA was tested under anaerobic conditions in three soil types: Massachusetts sandy loam (MSL), Arkansas silty loam (ASL), and California clay loam (CCL). For the composition of these soils see the paragraph above. TLC showed biodegradation of TBBPA in all soil types. The temperature during the study was 19-25 °C (mean 21.4 °C). Less than 0.5% of the applied radioactive TBBPA was recovered in the volatile traps, indicating only partial degradation. The recovered radioactivity in the traps was almost exclusively CO_2. Results of the TLC analysis indicated variable degradation rates of TBBPA. After 64 days, the amounts of TBBPA remaining in the soils were MSL: 43.7-57.4%, ASL: 53.4-65%, and CCL: 89.5-90.6%. Radioactivity recovered from the water ranged from 0.5 to 2.5% (Fackler, 1989d).

In another study, the biodegradability of ^{14}C-TBBPA was tested under aerobic conditions in a sediment/water microbial test system using natural river sediment and water. The test conditions were pH 5.5, field moisture capacity 15.9%, temperature 24-26 °C and the composition of the soil (6.8% carbon) was 92% sand, 6% silt, and 2% clay. Oxygen was bubbled through the system for 5 min/day to maintain aerobic conditions. The sampling intervals were scheduled on days 0, 4, 7, 10, 14, 21, 28, 42, and 56. Results from a 56-day aerobic test showed biodegradation of TBBPA in all tested concentrations, e.g., 0.01, 0.1, and 1 mg/litre. Half-lives calculated for TBBPA in the sediment/water microbial test systems ranged between 48 (10 μg/litre) and 84 days (1000 μg/litre), with apparent correlations between half-life and TBBPA concentration, and half-life and microbial population. The half-life in sterile soil could be extrapolated to be 1300 days, clearly indicating that the degradation observed in the active test systems was due to microbial degradation rather than physical processes. Less than 8% of the applied radioactive carbon from TBBPA was recovered in the volatile (CO_2) traps indicating only partial degradation. Filtered water contained less than 5% of the applied radioactivity. The amount of radioactivity observed to be remaining in the sediment at test termination, 44.7, 64.2, and 60.8% in the 0.01, 0.1 and 1 mg radioactive TBBPA/litre treatments, respectively, was comparable to the amounts reported in the aerobic degradation study in soil (Fackler, 1989e).

A biodegradation study on TBBPA (100 mg/litre) using sludge (30 mg/litre) for 2 weeks under sewage treatment condition showed no degradation by means of BOD (Chemical Inspection & Testing Institute, 1992).

4.2.3 Photodegradation

The calculated half-life of decomposition of TBBPA in water by UVR was 10.2 days in spring, 6.6 in summer, 25.9 in autumn, and 80.7 days in winter. Cloud cover lengthened the calculated half-life by a factor of 2. The water depth influenced the direct photodegradation more as the UV-absorption of the given body of water increased (Bayer, 1990).

In photodegradation experiments, TBBPA absorbed onto silica gel was exposed to UVR (254 nm). Eight metabolites were detected. The half-life value for TBBPA obtained in this test was 0.12 days (Bayer, 1990). It is difficult to derive environmental conclusions from the results of these experiments.

4.2.4 Bioaccumulation

TBBPA was labelled with ^{14}C in the aromatic ring. Blue gill sunfish (*Lepomis macrochirus*) (0.5-2.0 g) were exposed in a flow-through system for a period of 28 days to 0.0098 ± 0.0014 mg per litre. This was followed by a 14-day withdrawal period and bioaccumulation in edible tissue was determined. The average tissue concentrations of ^{14}C were found to be 0.196 mg/kg edible tissue and 1.69 mg/kg non-edible tissue. These values translated to bioconcentration factors (BCF) of 20 in edible tissue and 170 in visceral tissues. Plateau levels were reached within 3-7 days. In the fish, the whole body half-life was < 24 h. The radiocarbon dissipation to less than 0.01 mg/kg in the fish tissue occurred within 3-7 days of the beginning of the withdrawal phase (Nye, 1978).

TBBPA bioconcentration was determined in a 14-day toxicity study on *Chironomous tentans* (section 9.1.3). The bioconcentration factors calculated as the ratio of body concentration and interstitial water concentration ranged from 240 to 510 in high organic carbon sediments, 490 to 1100 in medium organic carbon sediments, and 650 to 3200 in low organic carbon sediments. Bioavailability of TBBPA increased with decreasing total organic carbon concentrations in the sediments (Breteler, 1989).

Fathead minnows (*Pimephalus promelas*) (20 fish) were continuously exposed to a mean measured concentration of 4.7 μg/litre, in a flow-through system, throughout the 24-day exposure period. The mean length and weight of the fish were 39 ± 4 mm and 0.57 ± 0.2 g, respectively. The water quality was:

hardness and alkalinity 28 and 20-26 mg/litre as $CaCO_3$, respectively, pH 7.0-7.6, dissolved oxygen 86-96% of saturation, and the temperature 19-21 °C. Throughout the 6-day depuration period, concentrations of ^{14}C remained below the limit of radiometric detection in water (0.29 μg/litre). The concentration of ^{14}C-residue in the tissue of fish reached a steady-state level on the fourth day of exposure. The mean steady-state tissue concentration was 5.8 mg/kg, which established a bioconcentration factor (BCF) of 1200. Following 6 days of depuration, 98% of the accumulated ^{14}C residues were eliminated from the tissues of exposed fish. The whole-body half-life was less than 1 day (Fackler, 1989c).

Eastern oysters (*Crassostrea virginica*) were continuously exposed to a mean, measured concentration of 1.0 μg/litre seawater for a 20-day exposure period. The valve height of the oysters ranged from 30 to 49 mm and they were determined to be immature by examination of the gonads. Seawater salinity was 32-34‰, pH 7.2-8.1, mean dissolved oxygen 7.4-7.5 mg/litre, and temperature 19 °C. Throughout the 14-day depuration period, concentrations of ^{14}C remained below the limit of radiometric detection in water (0.34 μg/litre). The concentration of ^{14}C residues reached a steady state level on the fifth day of exposure. The bioconcentration factor (BCF) was 780. The half-life of the ^{14}C residues in the oysters was between 3 and 5 days (Fackler, 1989b).

A bioaccumulation study on TBBPA (80 μg/litre, 8 μg/litre) using carp for 8 weeks showed 30-341 and 52-485 times bioaccumulation, respectively (Chemical Inspection & Testing Institute, 1992).

Regression equations were used to estimate the bioconcentration factor in fish using log P_{ow}. Using the value 4.48 of log P_{ow}, a log BCF of 3.2 is obtained. Since a substantial fraction of TBBPA is expected to be ionized and more polar at environmental pH values, and this fraction is less readily taken up by lipid membranes of the gill (depending on the counteracting influence of the bulky, non-polar bromine substituents, which may "mask" the ionized hydroxyl group), the amount of TBBPA in a form readily concentrated may be diminished. This probably accounts for the lower BCF values determined experimentally (Gustafsson & Wallen, 1988).

TBBPA has pKa values of 7.5 and 8.5. The aquatic toxicity tests were conducted at pHs ranging from 6.7 to 8.2. Interpretation of data for studies where the pH is close to the pKa may be difficult, because toxicity, bioaccumulation, depuration rates, and sediment binding will all be affected by the degree of dissociation exhibited. In addition, the behaviour of TBBPA in acidic waters may be different from that in the test situation.

4.3 Interaction with other physical, chemical, and biological factors

4.3.1 Pyrolysis

Purified TBBPA was pyrolysed in open quartz tubes at 700, 800, or 900 °C for 10 min. The residues were analysed for PBDD and PBDF. The pyrolysis of TBBPA gave mainly mono-, di-, tri-, and tetra-PBDD and -PBDF, but no highly brominated dibenzo-dioxins or dibenzofurans were found. The PBDD and PBDF formation was 0.02, 0.16, and 0.10%, respectively at 700, 800, and 900 °C. At 900 °C, the substance was partly decomposed. At 800 °C, the TeBDD and TeBDF isomers were produced at a concentration of 27 and 21 mg/kg of TBBPA, respectively (Thoma et al., 1986b).

Thies et al. (1990) pyrolysed pure TBBPA at 600 °C (10 or 20 min) producing primarily mono-tetra-BDF and BDD at individual concentrations of up to 130 000 µg/kg. 2,3,7,8-substituted congeners were produced at maximum concentrations of 10-50 µg/kg.

4.3.2 Pyrolysis of TBBPA-containing polymers

Epoxide resin with TBBPA, with 4-8% Sb_2O_3 or without Sb_2O_3, was tested for the formation of PBDD and PBDF by pyrolysis in a quartz tube at 400-800 °C, under aerobic conditions. Under these conditions, no PBDF or PBDD was found (limits of determination, 20 and 10 mg/kg, respectively) (Clausen et al., 1987).

The formation of 2,3,7,8-TeBBD and 2,3,7,8-TeBDF from epoxide resin with TBBPA was studied in pyrolysis experiments at 400, 600, and 800 °C. The following samples were studied; epoxide resin with 6% TBBPA in combination with 5% Sb_2O_3 and copper oxide (CuO); epoxide resin with 6% TBBPA and copper oxide and epoxide resin with 6% TBBPA and copper.

2,3,7,8-TeBDD was not detected at 400 °C, but, at 600 and 800 °C, in all three samples 1,3,6,8- and/or 1,3,7,9-tetrabromodibenzodioxin were found in concentrations of between 2.0 and 6.0 mg/kg. No 2,3,7,8-TeBDD or 2,3,7,8-TeBDF was found (limit of determination 0.01 mg/kg) (Lahaniatis et al., 1991).

Dumler et al. (1989) pyrolysed polymers (in granular form) mixed with TBBPA. The polymer mixtures were: Epoxide laminate/TBBPA; Epoxide laminate/TBBPA/copper laminate; PBT/TBBPA and Polycarbonate/TBBPA. Three different ovens were used; the DIN-oven, the BIS-oven, and the VCI-oven. The temperatures were 600 and 800 °C. Both the pyrolysis gases and the solid residues were analysed for PBDF and PBDD. PBDF was found in almost all samples. Polymers containing TBBPA generated small quantities of PBDF on pyrolysis, with yields ranging up to a few mg/kg. Mono- to tri-brominated congeners were identified.

The formation of PBDD and PBDF was studied during the pyrolysis of acrylonitrile/butadiene/styrene (ABS) with TBBPA at different temperatures and carrier gas compositions; mono- to pentabromodibenzofurans were formed at a μg/kg level. The optimum temperature of formation of PBDD and PBDF was 600 °C. The thermal degradation processes of the polymer were investigated in a thermogravimetric analysis. TBBPA did not exert any influence on the elementary chemical degradation processes of ABS. The flame retardant activity of TBBPA consists of the emission of brominated radicals and reduced flammability. The mechanism of formation of PBDD and PBDF from TBBPA, was found to be only a gas phase mechanism (Luijk & Govers, 1992).

Macro-pyrolysis experiments were performed in a quartz tube reactor. The ABS sample was inserted in the pre-heated tube and exposed at 400-700 °C for 20 min. The carrier gas was nitrogen, nitrogen with 5% oxygen, or, nitrogen with 10% oxygen. In a nitrogen atmosphere, predominantly mono- to pentabromodibenzofuran were formed at μg/kg levels. In the presence of oxygen, the yield of PBDF was increased. Although the formation of PBDD has been shown in an oxygen atmosphere, the yield of PBDD was lower than that of PBDF. At 600 °C, a maximum yield of both PBDD and PBDF was found. At 700 °C, a shift towards lower brominated compounds was observed. In neither of the samples were 2,3,7,8-substituted isomers detected. The results are summarized in Table 2 (Luijk & Govers, 1992).

Thies et al. (1990) pyrolysed commercial TBBPA-containing polymers at 600 °C under 3 different test conditions. Pyrolysis of polymer 1 (ABS/16% TBBPA/6% Sb_2O_3) produced a total of approximately 1500, 150, 3000 µg PBDD/DF/kg in the three tests, relative to the original sample weight. The pyrolysis products of two further polymers (TBBPA/bisphenol A - copolycarbonate (PC) + 10% copolymerized TBBPA and ABS/TBBPA/bisphenol A - polycarbonate blend + 6% copolymerized TBBPA) gave predominantly mono- to tribrominated PBDD and PBDF in the range of 100-5000 µg/kg. No 2,3,7,8-substituted isomers were detected (detection limits 1-4 µg/kg) with the exception of one sample, which showed 4 µg 2,3,7,8-TBDD/kg and 2 µg 2,3,7,8-TBDF/kg.

Table 2. The yields of PBDF and PBDD during pyrolysis of ABS/TBBPA in µg/kg relative to blend[a]

(a) PBDF

Temperature (° C)	MBDF[b]	DiBDF[b]	TrBDF[b]	TeBDF[b]	PeBDF[b]
			NITROGEN		
400	10 (3)	35 (25)	6 (4)	4 (2)	13 (2)
500	50 (43)	30 (7)	11 (3)	13 (3)	15 (5)
600	10	170	50	80	80
700	n.d.	840	50	50	n.d.
			NITROGEN + 5% OXYGEN		
400	10	25	3	10	10
500	5	40	15	50	50
600	10 (1)	925 (75)	200 (60)	100 (50)	50 (25)
700	200 (150)	2250 (650)	230 (30)	15 (4)	4 (2)
			NITROGEN + 10% OXYGEN		
400	n.d.	55	15	20	n.d.
500	20	190	15	15	20
600	265 (115)	1550 (1000)	220 (80)	70 (35)	35 (20)
700	130	2400	420	85	35

Table 2 (contd).

(*b*) PBDD

Temperature (° C)	MBDD	DiBDD	TrBDD	TeBDD	PeBDD
			NITROGEN		
400	n.d.	n.d.	0.05 (0.05)	n.d.	n.d.
500	n.d.	0.5 (0.5)	0.5 (0.1)	n.d.	n.d.
600	2	5	2	n.d.	n.d.
			NITROGEN + 5% OXYGEN		
400	1	1	1	n.d.	n.d.
500	3	15	15	n.d.	n.d.
600	5 (1)	250 (0)	145 (5)	3 (0.5)	n.d.
700	100 (80)	225 (75)	35 (5)	n.d.	n.d.
			NITROGEN + 10% OXYGEN		
400	n.d.	2	5	1	n.d.
500	6	70	40	2	n.d.
600	6 (1)	225 (155)	220 (0)	6 (0.1)	n.d.
700	5	75	30	n.d.	n.d.

[a] From: Luijk & Govers (1992).
[b] n.d. = not detected.

4.3.3 *Extrusion experiments with TBBPA-containing polymers*

A sample of ABS/16% TBBPA/6% Sb_2O_3 was heated to 240 °C for 20 min (Thies et al., 1990). In one test run, 100 μg/kg mono- and di-BDF were found in the post-extrusion resin, in the second run, less than 17 μg/kg were found. No 2,3,7,8-substituted PBDD/F were detected (detection limit 10 μg/kg).

Craig et al. (1989) carried out a study to determine whether brominated flame retardants and/or brominated dibenzodioxins and dibenzofurans were present in the fumes emitted during the thermal processing of resins. Thermal processing (heat treatment) involved extrusion of pelletized Cycolac resin formulated with, or without, TBBPA, under conditions considered to be representative of customer use. The measured die-zone and extrusion temperatures were 232 and 215 °C, respectively. A mass balance was obtained by analysing the residues in the pelletized extruder feed material (pre-extruded resin), in the heat-treated plastic resin (post-extruded resin), and in the fumes that were evolved during thermal processing. The results are summarized in Table 3.

Table 3. Total concentration of PBDD/PBDF (µg/kg)[a]

Flame retardant	Pre-extrusion resin		Post-extrusion resin		Fumes[b]	
	PBDF	PBDD	PBDF	PBDD	PBDF	PBDD
TBBPA	1.09 (1.09)[d]	n.d.[c]	n.d.[c]	6.16	0.020	0.006

[a] From: Craig et al. (1989).
[b] Fumes expressed as µg/kg of extruded resin.
[c] n.d. = Not detected (Detection limit 0.0002-0.075 µg/kg for PBDD and 0.002-0.205 µg/kg for PBDF).
[d] Total concentration of 2,3,7,8-substituted isomers.

Low levels of PBDF and PBDD were present in the pre- and post-extruded resin and in the fumes. The concentrations of identified 2,3,7,8-substituted isomers formed from TBBPA were just above the limit of detection.

4.3.4 Reports on fires involving TBBPA

A large-scale fire occurred in a storage area of a plastics production plant in Germany. Besides a great quantity of polycarbonate (PC) and polybutylene terephthalate (PBT), 180 tonnes of flame-retarded PBT were also burnt in the fire. The flame-retarded PBT contained TBBPA or its polymeric derivatives as a bromine carrier. Four samples of the mostly burnt PBT material and one sample of ash/slag mixture were examined for the presence of PBDF/PBDD residues. Three samples of soil were also analysed. One of the three soil samples was collected at a distance of 1460 m, one at 1340 m, and the third sample at a distance of 1740 m from the fire. The four samples of burnt PBT material and the ash/slag samples were analysed for the presence of 2,3,7,8-substituted tetra-, penta-, and hexa-BDF/BDD (eight congeners). The maximum concentrations detected were 0.5 μg/kg (detection limit 0.2-5 μg/kg). The three soil samples were analysed for the same congeners and concentrations of < 0.5 (detection limit) and 1.0 ng/kg were found (Neupert & Pump, 1992).

4.4 Ultimate fate following use

4.4.1 Disposal

It must be assumed that the majority of articles flame-retarded with TBBPA are ultimately disposed of either in landfills or incinerators.

4.4.2 Recycling of TBBPA-containing polymers

Studies on the reprocessability of selected samples of flame-retarded office machines have shown that those based on TBBPA can be recycled (Meyer et al. 1993). The materials tested were mainly flame-retarded ABS and PC-ABS polymer blends (most-used in office machines). Samples were tested from each of the following: granulated form, newly-produced parts, used parts with exact specification as to flame retardant, and two mixed used samples with unknown flame retardant from dismantled office machines. The samples were analysed for their contents of the 2,3,7,8-substituted congeners of polybrominated dibenzodioxins and furans (PBDD/PBDF) using ^{13}C internal standards of these isomers. Detection was by low resolution, and, for some samples, high resolution mass spectrometry. ABS samples with TBBPA contained only traces (less than 5 μg/kg) of these dioxin and furan isomers, even after 5 recyclings.

Lorenz & Bahadir (1993) investigated the recycling of printed circuits containing TBBPA, which is normally used in products of the German printed circuits industry. In the pilot recycling plant, a test run was carried out using a hammer mill and an impact grinder. No halogenated dibenzodioxins and dibenzofurans could be detected on the filters of active air sampling behind the filter devices of the mills. The shredded material was contaminated with PBDD/PBDF at low concentrations of 0.03-1.13 ng/g. In a test with printed circuits under thermal stress (up to 300 °C in an oven), low amounts of PBDD/PBDF (0.74-4.52 ng/g) were generated. The authors concluded that printed circuits containing TBBPA can be recycled.

5. ENVIRONMENTAL LEVELS AND HUMAN EXPOSURE

5.1 Environmental levels

5.1.1 Air

The air near production facilities in Southern Arkansas, USA, contained 1.8 μg TBBPA/m^3 (Zweidinger et al., 1979).

5.1.2 Water

In 1977, in Japan, none of 15 water samples analysed contained TBBPA (limit of determination, 0.02-0.04 μg/litre). Water samples were collected in 25 areas in Japan in 1986-87. TBBPA was detected in one out of three water samples from the mouth of the Yamato River (Environment Agency Japan, 1989). In 1987, in Japan, TBBPA was detected in one out of 75 water samples at a concentration of 0.05 μg/litre (limit of determination, 0.03 μg/litre). In 1988-89, in Japan, TBBPA was not detected in 150 water samples collected at 50 locations (limit of determination, 0.04 μg/litre) (Environment Agency Japan, 1989, 1991).

5.1.3 Soil

TBBPA was detected at 0.5-140 μg/kg (dry weight) in 14 out of 19 river sediment samples in Osaka, Japan. In marine sediments in Osaka bay, levels of 0.5-4.5 μg/kg (dry weight) were found in 1981-83. In the marine sediments of two areas other than Osaka, the levels were much lower (n.d.-1.8 μg/kg dry weight). The dimethylated metabolite of TBBPA was found in 5 out of 6 samples of river sediment, collected in the Osaka area in 1983, in concentrations of 0.6-1.8 μg/kg wet weight (Watanabe et al., 1983a,b; Watanabe & Tatsukawa, 1990). A river sediment was collected in 1981 downstream of the Neya River, from a tributary of the Yodo River, which empties into the Osaka Bay. The TBBPA concentration was about 20 μg/kg dry weight (Watanabe et al., 1983a,b).

Sediment samples were collected in 22 areas, in Japan, in 1987. TBBPA was found in the bottom sediments from 6 areas; the mouth of the Sumida River (3/3), the mouth of Ara River (1/3), the mouth of Yamato River (3/3), the river flowing in Osaka City (3/3), the Port of Osaka (3/3) and the mouth of Yodo River (1/3).

TBBPA was detected in 14 out of 66 sediment samples at concentrations ranging from 2 to 150 μg/kg dry weight and it was detected in 20 out of 130 sediment samples collected at 44 locations at concentrations ranging from 2 to 108 µg/kg dry weight in the 1988 (limit of determination in both studies: 2 µg/kg dry weight) (Environment Agency Japan, 1989, 1991).

Sellström et al. (1990) analysed sediment samples taken upstream and downstream from a factory in Sweden for the presence of TBBPA and its dimethylated derivative (Me_2-TBBPA). The downstream level of TBBPA was 430 µg/kg (ign. loss) and upstream, 50 µg/kg, the levels of the dimethylated compound were 2400 µg/kg (ign. loss) and 36 µg/kg, respectively.

5.1.4 Fish and shellfish

TBBPA was not detected in mussels (*Mytilus edulis*) collected in Osaka bay in 1981. Two out of 19 samples of fish and shellfish, collected in the Osaka area, contained 0.8 and 4.6 µg methylated TBBPA/kg wet weight, respectively (Watanabe & Tatsukawa, 1990).

When 75 fish samples were collected in 24 areas in Japan in 1987, no TBBPA was detected (Environment Agency Japan, 1989). TBBPA was also not detected in 135 samples collected at 45 different locations in Japan in 1988 (limit of determination, 1 µg/kg wet weight) (Environment Agency Japan, 1991).

5.2 General population exposure

The dimethylated metabolite of TBBPA was not found in 5 fat samples collected from people living in the Osaka area (limit of determination, < 20 µg/kg fat) (Watanabe & Tatsukawa, 1990).

5.3 Occupational exposure

There are no data on TBBPA occupational exposure levels.

Investigations into the possible formation of PBDD/PBDF during processing have been described in section 4.3.3.

Thies et al. (1990) monitored the workplace atmosphere near the system for manually controlling an injection moulding machine, when processing a polymer formulation (ABS + 16% TBBPA/6% antimony trioxide). Two samples were taken at 4.3 and 4.8 m^3. At detection limits of 0.1 and 1 ng/m^3 respectively, no 2,3,7,8-PBDD/PBDF isomers or PBDD/PBDF were detectable.

6. KINETICS AND METABOLISM IN LABORATORY ANIMALS AND HUMANS

6.1 Absorption and elimination

6.1.1 Mammals

A single oral dose (6.51-7.55 mg/kg body weight) of ^{14}C TBBPA (phenyl UL-^{14}C) in corn oil was poorly absorbed from the gastrointestinal tract of Sprague-Dawley rats. About 95% of ^{14}C-labelled material was eliminated in the faeces and less than 1.1% in the urine within 72 h. The highest tissue levels were found in the liver and in the gonads. Tissue half-lives were reported to be 19.9 h in the blood, 70.8 h in fat, 17.1 h in the kidneys, 10.8 h in the liver, 39.3 h in the spleen, 48.0 h in muscle, and 60.5 h in the gonads. The maximum half-life in any tissue is less than 3 days (Brady, 1979).

6.1.2 Fish and shell-fish

When bluegill sunfish (*Lepomis macrochirus*) were exposed to labelled ^{14}C-TBBPA in the water at a concentration of 0.0098 mg/litre, they showed rapid uptake of TBBPA. Equilibrium was reached within 3 days. The ^{14}C in the fish was rapidly eliminated on transfer to uncontaminated water. The half-life of elimination was less than 24 h in both edible and non-edible tissues. The residues decreased below the limit of detection (< 0.01 mg/kg) in 3-7 days (Nye, 1978) (see also section 4.2.4).

TBBPA was not detected in mussels (*Mytilus edulis*) collected at the seashore in Osaka Bay in 1981. However, a 4.4′-dimethoxy derivative of TBBPA was identified in the mussel at a concentration of 5 µg/kg wet weight (Watanabe et al., 1983a).

6.2 Metabolism

No data are available.

7. EFFECTS ON LABORATORY MAMMALS AND *IN VITRO* TEST SYSTEMS

7.1 Single exposure

7.1.1 Oral

Gustafsson & Wallen (1988) reported oral LD_{50}s of > 2 g TBBPA/kg body weight in rats and 3.2 g/kg body weight in mice.

TBBPA was administered to a group of 10 rats (5 male, 5 female) at a single dose of 5000 mg/kg. The animals were observed for 14 days. No mortality occurred during the observation period. No gross lesions were detected at necropsy. The rat oral LD_{50} was > 5000 mg/kg (Hardy, 1994).

The LD_{50} in B6C3F1 mice was reported to be 4.4 g/kg and 4.5 g/kg in male and female mice, respectively (Sekizawa, personal communication, 1994).

7.1.2 Dermal

TBBPA was applied to the clipped, intact skin of albino rabbits at concentrations of up to 3.16 g/kg body weight, for 24 h. No local or systemic symptoms could be detected by clinical observation, urinalysis, haematology, weight gain, or gross pathology (Great Lakes Chemical Corporation, 1986, summary report).

The LD_{50} of TBBPA in guinea-pigs was > 1 g/kg body weight (Bayer, 1990, summary report).

TBBPA was applied at a dose level of 2000 mg/kg to 10 rabbits (5 male, 5 female). The test material was applied on abraded skin, covered, and left in contact for 24 h. The animals were observed for 14 days. No mortality occurred during the observation period. Slight erythema and oedema were observed in 1 rabbit on day 1. No gross lesions were detected at necropsy. The rabbit dermal LD_{50} is > 2000 mg/kg (Hardy, 1994).

When 10 female rabbits were dermally exposed to TBBPA at a dose of 200 mg/kg body weight, all rabbits exhibited reddening of the skin that returned to normal in 48 h. None of the rabbits died. The LD_{50} was > 200 mg/kg. In another study, the intact or shaven skin of groups of 2 rabbits was exposed to a dose of 1, 2.15, 4.64, or 10 g/kg body weight for 24 h. Weight loss was seen at the two highest dose levels. Only one rabbit died in the 1 g/kg group and one in the 4.64 g/kg group (Bayer, 1990 - summary report).

7.1.3 *Inhalation*

Groups of 10 Wistar rats, 10 NMDI-mice, and 10 guinea-pigs (five of each sex/group) were exposed for 8 h to a concentration of 0.5 mg TBBPA aerosol/litre air and observed for 48 h after exposure. None of the animals showed symptoms of local or systemic toxicity. There were no gross pathological findings at autopsy (Sterner, 1967).

7.2 Short-term exposures

7.2.1 *Oral (rat)*

Groups of 25 female and 25 male Charles River CD rats (males 260-341 g, females 183-232 g) were fed dietary levels of 0, 1, 10, 100, or 1000 mg TBBPA/kg (corresponding to 0, 0.05, 0.5, 5, or 50 mg/kg body weight per day) for 28 days. After 4 weeks, 5 rats/sex per group were sacrificed and the remaining rats placed on control diets for 2, 6, or 12 weeks. No effects on behaviour, appearance, food consumption, body weight gain, or mortality were observed. No gross or microscopic abnormalities were observed. Total bromine levels were determined in the liver and fat of rats from the control group and the highest dose group sacrificed at the end of the 28-day feeding period. No differences in bromine contents were seen (Goldenthal & Geil, 1972).

Groups of 7 male and 7 female Sprague-Dawley rats (6-7 weeks old), (control and 3 mg/kg groups consisted of 21 male and 21 female animals) were fed a diet supplying 0, 0.3, 3, 30, or 100 mg TBBPA/kg body weight for 90 days. The concentrations in the diet were adjusted weekly to deliver the above doses. The toxicological parameters evaluated were appearance, demeanour, body weight gain, food consumption, haematology, clinical chemical determinations, urinalysis, organ weights, and gross- and microscopic examinations. The administration of TBBPA in the diet at a dose as high as 100 mg/kg body weight per day for 90 days did not produce toxicological effects. On days 10, 20, 30, 60 and 90, liver, kidneys, skeletal muscle, fat and serum of 2 control animals and 2 animals of the 3 mg/kg group were analysed for bromine. The total bromine content in the tissues of rats receiving 3 mg/kg per day did not differ from that of the controls. Higher dose levels were not tested (Quast et al., 1975).

In a Japanese study (Tobe et al., 1986), B6C3F1 mice (10/sex per group) were fed TBBPA in the diet at 0, 500, 4900, 15 600, or 50 000 mg/kg (corresponding to 0, 71, 700, 2200, or 7100 mg/kg body weight per day, respectively) for 3 months. All animals at the highest dose died during the study, probably because of malnutrition and anaemia. No deaths were observed at lower doses. Body weight gains were decreased at levels of 15 600 mg/kg and higher, though food intake did not change. Red blood cells, haemoglobin, haematocrit, serum triglycerides, and total serum proteins decreased at 15 600 mg/kg. Organ weight changes and pathological changes were not detected, except in the spleen, where organ weight increased and some blood was observed outside the medulla. These effects may have been related to the haemorrhage observed in this study and the uncoupling of energy production in mitochondria observed by Inouye et al. (1979). The NOAEL was 4900 mg/kg diet (corresponding to 700 mg/kg body weight per day).

7.2.2 *Inhalation (rat)*

Four groups of 5 female and 5 male Charles River CD rats (males 260-334 g, females 213-248 g) were exposed to an atmosphere of 0, 2, 6, or 18 mg micronized TBBPA/litre air (0, 2000, 6000, or 18 000 mg/m^3) for 4 h daily, 5 days/week, for 2 weeks. Body weights and food consumption were recorded weekly. Haematological and biochemical examinations and urinalysis were carried out just before the rats were sacrificed. Excessive salivation, red or clear nasal discharge, and excessive lacrimation were noted during the course of the study for rats at the two highest dose levels. There were no deaths and no changes in body weight gain, food consumption, haematological and biochemical parameters, or urinalysis. A decrease in relative liver weight of the female animals from the three dose levels might have been compound related. No gross or microscopic lesions were seen in any of the rats at the end of the study (Goldenthal et al., 1975).

7.2.3 *Dermal (rabbit)*

Dosages of 100, 500, or 2500 mg TBBPA/kg body weight were applied to the backs of New Zealand white rabbits (2030-2311 g), for 6 h/day, 5 days a week, for 3 weeks. Four male and four female rabbits were used at each dose level and also in the control group. The control group received 0.9% physiological saline. The back of each rabbit was clipped with an electric clipper and the skin of one-half of the rabbits in each group was abraded twice each week. No mortality or signs of overt toxicity were observed. Very slight skin erythema was observed occasionally at the low dose and, for almost all rabbits for various lengths of time, at the two higher dose levels. Body weight gain, haematological parameters, urinalysis, organ weights, and gross and microscopic examinations did not reveal any compound-related changes (Goldenthal et al., 1979).

7.3 Long-term exposure

No data on this subject are available.

7.4 Skin and eye irritation; sensitization

7.4.1 Skin irritation

The studies conducted in which rats were administered TBBPA did not reveal skin irritation (Quast et al., 1975).

Three male and three female rabbits were administered 500 mg TBBPA on the intact and shaven skin for 24 h. No skin irritation was observed. In another study, six rabbits were exposed to 500 mg on intact and shaven skin for 24 h, no deaths occurred (Bayer, 1990, summary report).

TBBPA was applied to 2 intact and 2 abraded skin sites on each of 6 rabbits (3 males, 3 females). The application site was covered for 24 h. Observations were recorded 24 and 72 h after exposure. Scores for all animals for all readings were zero. TBBPA was non-irritating to the skin (Hardy, 1994).

7.4.2 Eye irritation

TBBPA was instilled into the right eye of six rabbits (3 males, 3 females). Four of the rabbits exhibited slight redness at the 1-h observation period (four scores of Grade 1 in the conjunctiva). No other ocular reactions were observed during the study. TBBPA was non-irritating to the eye (Hardy, 1994).

Single doses of 3 mg of finely ground TBBPA were applied to the conjunctival sac of the eye of New Zealand White rabbits

(2.5 kg). Eye examinations were carried out, 5 min, and 1 and 4 h after the application, and daily thereafter for 7 days. No effects on the cornea, iris, or conjunctiva using Draize score were observed at any time. The rabbits exhibited normal appearance and behaviour, gained weight normally, and showed no evidence of systemic toxicity. There were no gross pathological findings at autopsy. It was concluded that TBBPA was not irritating to the eye (Sterner, 1967).

7.4.3 *Sensitization*

Twelve male, albino guinea-pigs (596-700 g) were divided into two groups consisting of a positive control group (dinitrochlorobenzene) of 4 guinea-pigs and a treated group of 8 guinea-pigs. The compounds were injected intradermally every other day, 3 times/week, up to a total of 10 doses in the back and flanks of the guinea-pigs. TBBPA was applied at a concentration of 0.1% TBBPA in 0.9% NaCl solution. The solvent was applied to all animals in the other flanks. Two weeks after the 10 injections, a challenge dose was administered intradermally. The result of the TBBPA challenge injection was negative. No sensitization reaction was found (Dean et al., 1978b).

TBBPA was applied dermally to 10 guinea-pigs for a total of 9, six-h insult periods. A positive control group consisting of 10 guinea-pigs was treated with 2,4-dinitrochlorobenzene. Approximately 14 days after the last sensitizing exposure, the animals were challenged in the same manner at both the site of sensitization and a second site. A second challenge was made 48 h after the first challenge. A positive response was elicited by the positive control substance. No irritation was observed during induction or challenge with TBBPA. TBBPA was not a sensitizer in this guinea-pig sensitization test (Hardy, 1994).

7.4.4 *Chloracnegenic activity*

TBBPA was inuncted in one ear of each of 4 rabbits (2 males and 2 females) at concentrations of 0.5, 5, or 50% in Polylan. The substance was administered once daily, 5 days/week, for 4 weeks. Observations were recorded at time 0, and on days 7, 14, 21, and 28. No positive scores were recorded for concentrations of 5 and 50%. One rabbit exhibited a slight response (grade 1) at the 0.5% concentration on day 7. No other positive reactions were observed. No gross lesions were recorded at necropsy. TBBPA was noncomedogenic in the rabbit ear assay (Hardy, 1994).

7.5 Reproductive toxicity, embryotoxicity, and teratogenicity

7.5.1 Teratogenicity

TBBPA was administered, by gavage, at dose levels of 0, 30, 100, 300, 1000, 3000, or 10 000 mg/kg body weight, on gestation days 6-15, to groups of 5 Charles River CD female rats (15 weeks old). The rats were sacrificed on gestation day 20. Three out of 5 rats given 10 000 mg/kg died, while the remaining rats in this group showed a slight decrease in body weight gain between gestation days 6 and 15; green, soft stools; and an increase in matted hair in the anogenital area. There were no signs of toxicity in rats administered levels up to, and including, 3000 mg/kg. There were no differences in the mean numbers of viable or nonviable fetuses, resorptions, implantations, or corpora lutea compared with the controls (Goldenthal et al., 1978).

In another study, rats were treated with 0, 0.28, 0.83, or 2.5 g TBBPA/kg body weight from day 0 to day 19 of gestation. The treatments did not impair the birth rate. No toxic effects were observed on the embryo or fetus, and there were no skeletal or visceral abnormalities. The postnatal development was not impaired (Noda, 1985).

7.6 Mutagenicity and related end-points

A mutagenicity study was conducted on *Salmonella typhimurium* TA1535, TA1537, TA1538, TA98 and TA100 and on *Saccharomyces cerevisiae* strain D4, with, and without, metabolic activation with liver S9 fraction of Aroclor 1254-induced male Sprague-Dawley rats. Positive controls were used for comparison. The dose levels of TBBPA were 0, 0.25, 0.5, 5.0, and 50 μg/plate in DMSO. No mutagenic activity was found in this assay (Brusick, 1976).

TBBPA was examined for mutagenic activity at concentrations of 0.001, 0.003, 0.01, 0.03, and 0.1 mg/plate in a series of *in vitro* microbial assays using *Salmonella typhimurium* TA1535, TA1537, TA1538, TA98, and TA100 and *Saccharomyces cerevisiae* with, and without, microsomal enzyme preparations from Aroclor-induced rats. All tests with, and without, the liver activation system were negative (abstract only) (Great Lakes Chemical Corporation, 1986).

Mortelmans et al. (1986) tested TBBPA for mutagenic potential in *Salmonella typhimurium* strains TA100, TA1535, TA1537, and TA98 in concentrations of 0, 100, 333, 1000, 3333, and 10 000 μg/plate with, and without, S9 mix of Aroclor 1254-treated, male Sprague-Dawley rats and male Syrian hamsters. The substance was dissolved in DMSO. TBBPA did not show mutagenic potential.

Ethyl Corporation also reported that TBBPA was negative in several Ames assays in 5 strains both with and without exogenous metabolic activation (Hardy, 1994).

7.7 Carcinogenicity

No data are available on this subject.

7.8 Other special studies

The *in vitro* effect of TBBPA on the function of biological membranes was examined. Human erythrocytes or rat mitochondria were tested with TBBPA at 25-250 μmol/litre. The data indicate that TBBPA primarily alters the permeability of membranes, resulting in haemolysis of erythrocytes accompanied by morphological changes and uncoupling of the mitochondrial oxidative phosphorylation (Inouye et al., 1979).

In rats, after a single dose (oral or ip) of TBBPA, moderate microsomal enzyme-inducing activity was observed in the liver but not in the small intestine (Gustafsson & Wallen, 1988).

8. EFFECTS ON HUMANS

TBBPA mixed with water to produce a thick slurry of a paste-like consistency (approximately 3-5 mg) was applied 10 times to the upper arms of 13 male and 41 female volunteers during the sensitization phase. A modified Draize multiple insult test was conducted which was followed after 10-14 days by the challenge treatment. TBBPA did not produce any skin irritation and did not show any evidence of contact sensitization in the subjects who completed the study (Dean et al., 1978a).

9. EFFECTS ON OTHER ORGANISMS IN THE LABORATORY AND FIELD

9.1 Laboratory studies

It should be noted that results of studies performed at pHs close to the pKa values (7.5 and 8.5, respectively) may be difficult to extrapolate outside this range.

9.1.1 Microorganisms

9.1.1.1 Water

Marine unicellular algae, *Skeletonema costatum*, *Thalassiosira pseudonana*, and *Chlorella* sp., were exposed to TBBPA in 6 algal growth media. The duration of the exposure for *S. costatum* and *T. pseudonana* was 72 h, and, for *Chlorella* sp., 96 h. The population density was estimated by cell counts on a haemacytometer. Growth of *Chlorella* sp. was not inhibited by as much as 50% by 1500 µg/litre of TBBPA. TBBPA was toxic for *S. costatum* and *T. pseudonana*, the EC_{50}s being between 90-890 µg/litre and 130-1000 µg/litre, respectively (Walsh et al., 1987).

The freshwater green alga, *Selenastrum capricornutum* (5-day-old inoculum) was used to test TBBPA (prepared with distilled deionized water) at measured concentrations of 0.34, 0.76, 1.5, 3.0, and 5.6 mg/litre (nominal concentrations 0.6, 1.2, 2.4, 4.8, and 9.6 mg/litre)[a]. The conditions of the test were, temperature 20-24 °C, constant illumination, shaking at 100 rpm, and pH 7.5. The effect criterion was reduction in cell density relative to the control. Growth of *S. capricornutum* was not reduced by 96 h of exposure to TBBPA at any dose level (Giddings, 1988).

9.1.1.2 Soil

TBBPA was tested in two strains of bacteria capable of carrying out the *O*-methylation of phenolic compounds. The strains were; strain 1395 (Gram-positive *Rhodococcus* sp.) and strain 1678 (Gram-negative *Acinetobacter* sp.). Three ml of cell suspension was used in a 28-ml serum bottle. TBBPA was

---

[a] 0.72-4.16 mg/litre reported for the water solubility, but it may be pH-dependent.

O-methylated only by the Gram-positive strain. It was suggested that, in the natural environment, bacterial *O*-methylation of phenols carrying electron-attracting substituents might be a significant alternative for biodegradation (Allard et al., 1987).

9.1.2 Aquatic organisms

9.1.2.1 Invertebrates

The 48-h, acute LC_{50} for *Daphnia magna* (less than 20 h old) was 0.96 mg TBBPA/litre; at the lowest concentration studied (0.32 mg/litre), 5% of the organisms died. The water conditions were: temperature 17.5 °C, pH 7.32, and total hardness and total alkalinity, 64 and 32 mg/litre, respectively (Morrissey, 1978).

Daphnia magna were continuously exposed (flow-through system) for 21 days to measured concentrations of 0.056, 0.10, 0.19, 0.30, and 0.98 mg TBBPA (99.15%)/litre. Well water was used with a total hardness of 170 mg/litre and an alkalinity of 120 mg/litre (both as $CaCO_3$), pH 8.1-8.2, temperature 20 °C, and dissolved oxygen of 8.0-8.7 mg/litre. At the termination of the study, daphnid survival at all concentrations ranged from 95 to 100%, which was comparable with the 98% survival of control organisms. Daphnid growth, as determined by the measurement of individual body lengths at the end of the test, was also not adversely affected by any of the test concentrations. Reproduction, as determined by cumulative numbers of offspring per female at test termination, was the most sensitive indicator of toxicity of TBBPA for *Daphnia magna* in the concentration range tested. Reproduction in 0.98 mg/litre was 21 offspring per female, which was significantly less than the reproduction of the pooled control organisms (60 offspring per female). Reproduction at the remaining test concentrations was statistically similar to that of the pooled control organisms. The Maximum Acceptable Toxicant Concentration (MATC) for *Daphnia magna* was > 0.30 and < 0.98 mg/litre (geometric mean 0.54 mg/litre) (Surprenant, 1989b).

Steinberg et al. (1992) showed that dissolved humic material had no effect on the toxicity of TBBPA for *Daphnia magna* (inhibition of motility; 48 h; 20 °C).

Goodman et al. (1988) exposed Mysid shrimp (*Mysidopsis bahia*), aged < 1, 5, and 10 days old, to TBBPA in a flow-through system for 96 h. The test conditions were: mean salinity 20.6‰, pH 7.96-8.16, and mean dissolved oxygen concentration 6.9 mg/litre. The TBBPA was dissolved in a mixture of triethylene glycol and acetone. The 96-h LC_{50} values for the three live stages were 860(670-1200), 1100, and 1200 μg TBBPA/litre, respectively.

The acute EC_{50}, defined as reduction of shell deposition, was determined in Eastern oysters (*Crassostrea virginica*) in a flow-through system (oysters had a mean valve height of 41 mm). The mean measured test concentrations were 0.018, 0.032, 0.051, 0.087, and 0.150 mg TBBPA/litre. Salinity range was 29-32‰, dissolved oxygen ranged from 86 to 95% of saturation, pH 8.0-8.1. The 96-h EC_{50} was calculated to be 0.098 mg TBBPA/litre with a no-observed-effect concentration below 0.018 mg/litre. An estimated NOEC of 0.0062 mg/litre was calculated (Surprenant, 1989c).

9.1.2.2 *Fish*

The 96-h, acute LC_{50} of TBBPA for bluegill sunfish (*Lepomis macrochirus*; 6 months old, length 38 mm, weight 0.59 g) was 0.51 mg/litre (nominal concentration), in a static system. The conditions of the water were: temperature 21.7 °C, pH 7.47, and total hardness and total alkalinity, 44 mg/litre and 33 mg/litre as $CaCO_3$, respectively. With dose levels above 0.32 mg/litre, the fish became irritated and exhibited abnormal sounding and skittering swimming behaviour. The no-effect level was 0.10 mg/litre (Calmbacher, 1978a).

The 96-h LC_{50} of TBBPA for rainbow trout (*Salmo gairdneri*; 3 months old, length 41 mm, weight 0.51 g) was 0.40 mg/litre (nominal concentration) in a static system. The conditions of the water were: temperature 12.3 °C, pH 7.48, total hardness and total alkalinity, 40 and 35 mg/litre as $CaCO_3$, respectively). The no-effect level was 0.18 mg/litre. With higher levels, the fish became irritated and exhibited twitching, erratic swimming, dark discoloration, and laboured respiration (Calmbacher, 1978b).

The LC_{50} of TBBPA for fathead minnow (*Pimephales promelas*; mean wet weight 0.50 g and total length 36 mm) (20 fish/group) was determined under flow-through conditions. The total duration of the study was 144 h. Well-water was used with total hardness and alkalinity ranges of 22-30 and 21-24 mg/litre, as $CaCO_3$, respectively. The pH was 6.7-7.1, the dissolved oxygen concentration range, 91-96% of saturation, and the temperature 21-22 °C. The mean measured test concentrations were 0.19, 0.26, 0.32, 0.45, and 0.63 mg/litre. The 96-h LC_{50} was determined to be

0.54 mg/litre with a no-observed-effect concentration of 0.26 mg/litre (Surprenant, 1988).

Fathead minnow (*Pimephales promelas*) embryos and larvae were continuously exposed for 35 days (30 days post-hatch) to mean, measured TBBPA concentrations ranging from 0.024 to 0.31 mg/litre. The water quality was: mean total hardness 28-29 mg/litre and alkalinity 23-24 mg/litre (both as $CaCO_3$), pH 7.0-8.2, temperature 24 °C, and mean dissolved oxygen 8.1-8.6 mg/litre. Observations were made on the survival of organisms at hatch, and survival and growth (wet weight and total length) of larvae after 30 days post-hatch exposure. The survival at the end of the hatching period (day 5) at the highest concentration of 0.31 mg/litre was 28% and was significantly less than survival in the control organisms, 84% (pooled control and solvent control data). The survival of embryos exposed to mean concentrations of 0.16, 0.084, 0.040, and 0.024 mg/litre ranged from 74 to 90% and was unaffected compared with the control embryos. All larvae exposed to 0.31 mg/litre died within the initial 7 days of the post-hatch exposure period. The survival of larvae exposed to the remaining concentrations of TBBPA (0.16-0.024 mg/litre) ranged from 87 to 93% and was comparable to survival in the control larvae (93%). At test termination (30 days post-hatch), growth data (total length and wet weight) established that surviving fish at all treatment levels grew at rates comparable to those of the control larvae. The mean length and wet weight of larvae exposed to the mean, measured TBBPA concentration of 0.16 mg/litre ranged from 24 to 25 mm and 112 to 126 mg, respectively, and were statistically comparable to those of control larvae (pooled data, 25 mm and 111 mg, respectively). On the basis of adverse effects on embryo and larval survival, the Maximum Acceptable Toxicant Concentration (MATC) of TBBPA for fathead minnow was estimated to be > 0.16 mg/litre and < 0.31 mg/litre (geometric mean 0.22 mg/litre) (Surprenant, 1989a).

9.1.3 Sediment-dwelling organisms

A study with a benthic invertebrate midge, *Chironomous tentans* (25 per replicate vessel) consisted of three, 14-day (partial life cycle) toxicity tests under flow-through conditions. Each of the sediment tests was conducted with sediment containing different levels of organic carbon. The water quality was: mean total hardness and total alkalinity 29-30 and 25-28 mg/litre as $CaCO_3$, pH 6.9-7.8, temperature 22 °C, and dissolved oxygen

7.7-8.6 mg/litre. These values were slightly different in the different tests, depending on the quantity of organic matter. At the termination of the 14-day sediment studies, midge survival in all TBBPA-treated sediments ranged from 44 to 96% and was statistically comparable to the survival of controls. Organism growth (determined by the measurement of grouped body weights) at test termination was not significantly different for any of the 3 different levels of organic carbon. The high organic carbon (HOC), medium organic carbon (MOC), and low organic carbon (LOC) contents of the sediments were 68, 27, and 2.5 g/kg, respectively. The sediments were physically characterized by a high sand content of 920-940 g/kg, a silt content of 10-60 g/kg, and a clay content of 20-60 g/kg and were slightly acidic (pH 5.4-5.5). The mean, measured concentrations of TBBPA in HOC, MOC, and LOC, were 0.0044-0.046, 0.0075-0.045, and 0.0078-0.046 mg/litre, respectively. The highest no-effect level was established at an interstitial water concentration of 0.046 mg TBBPA/litre, which was the highest concentration attained in the HOC treatment. The TBBPA concentration in the HOC sediment was 340 mg/kg. The no-effect level in the interstitial waters of MOC and LOC treatments were 0.045 and 0.046 mg TBBPA/litre. The TBBPA concentrations of the sediments in MOC and LOC treatments were 240 and 230 mg/kg. Bioconcentration factors in the midge ranged from 240 to 510 in the HOC sediments, 490 to 1100 in the MOC sediments, and 650 to 3200 in the LOC sediments. A high organic content in the sediment reduced accumulation. No adverse biological effects resulted from the increased TBBPA body burden. No relationship was observed between the sediment concentration of TBBPA and midge body burden (section 4.2.4) (Breteler, 1989).

9.2 Field observations

No data are available on this subject.

9.3 Miscellaneous

Kawamura et al. (1986) designed a study to investigate the effects of TBBPA on the aerobic metabolism of the bisphenolic derivative-sensitive protozoon, *Giardia lamblia*. In this study, trophozoites of *G. lamblia* were grown in a medium for 72 h at 35.5 °C. The parasites, which were harvested and washed, and, finally, suspended in a buffered sucrose in a final protein concentration of 5-8 mg/ml, were disrupted by homogenization for 10 min. Inhibition of endogenous respiration, and the activities of NADH- and NADPH oxidase in *Giardia* by TBBPA were measured. The concentrations of TBBPA needed for 50% inhibition of endogenous respiration, NADH oxidase, and NADPH oxidase, were 0.30, 0.15, and 0.15 mmol/litre, respectively.

TETRABROMOBISPHENOL A DERIVATIVES

Five derivatives of TBBPA were identified as being in commercial use as flame retardants. These are: tetrabromobisphenol A dibromopropylether, tetrabromobisphenol A bis(allylether), tetrabromobisphenol A bis(2-hydroxyethyl ether), tetrabromobisphenol A carbonate oligomers, and tetrabromobisphenol A brominated epoxy oligomer. In addition, the dimethylated derivative of TBBPA has been identified in a few environmental samples. This dimethylated-TBBPA derivative is suspected of being an environmental metabolite of TBBPA.

Few data are available on these TBBPA derivatives. The available data are summarized in Appendix A. No evaluations or recommendations were made because of the lack of data. The five TBBPA-derived flame retardants are not extensively used (approximately 25% the global volume of TBBPA). They are believed to be used in specialized (or niche) applications.

A. TETRABROMOBISPHENOL A DIMETHYLETHER

A.1 SUMMARY AND EVALUATION; CONCLUSIONS AND RECOMMENDATIONS

There is no data base on which to make an evaluation of tetrabromobisphenol A dimethylether, or to support its use commercially.

Tetrabromobisphenol A dimethylether cannot be evaluated unless adequate data become available on physical and chemical properties, production and use, environmental transport, distribution, and transformation, environmental levels and human exposure, kinetics and metabolism in animals and humans, effects on laboratory mammals, humans, and *in vitro* test systems, and effects on other organisms in the laboratory and field.

A.2 IDENTITY, PHYSICAL AND CHEMICAL PROPERTIES, AND ANALYTICAL METHODS

A.2.1 Identity

Chemical formula	$C_{17}H_{16}Br_4O_2$
Chemical structure	

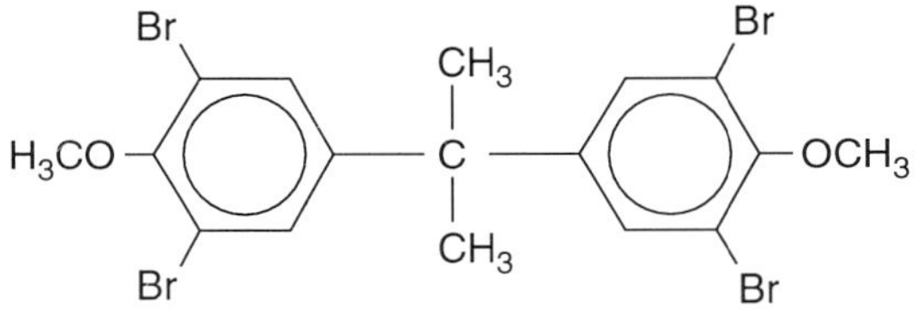

CAS registry number	37853-61-5
Synonyms	1,1′-(1-methylethylidene) bis(3,5-dibromo-4-methoxy)benzene;tetrabromobiphenyl A-bis(methylether); tetrabromobisphenyl A methylether
Relative molecular mass	571.9
Vapour pressure at 25 °C	2 x 10^{-7} Torr (Watanabe & Tatsukawa, 1990)
Log P_{ow}	6.4-7.6 (Watanabe & Tatsukawa, 1990; Sellström et al., 1994).

A.3 SOURCES OF HUMAN AND ENVIRONMENTAL EXPOSURE

As far as is known, the dimethylether of TBBPA is not used commercially as a flame retardant (McAllister, 1994, personal communication).

No data are available on environmental transport, distribution, and transformation.

A.4 ENVIRONMENTAL LEVELS AND HUMAN EXPOSURE

No data are available on the following subjects:

- Kinetics and metabolism in laboratory animals and humans
- Effects on laboratory mammals and *in vitro* test systems
- Effects on humans
- Effects on other organisms in the laboratory and field

A.4.1 Sediment

Me_2-TBBPA has been found in 5 out of 19 sediment samples from Japan at levels of 0.6-1.8 μg/kg dry weight (Watanabe & Tatsukawa, 1990). The dimethylated derivative was also found in Swedish sediments close to an industry using TBBPA. The level upstream from the factory was 36 μg/kg, and that downstream, 430 μg/kg ign.loss (Sellström et al., 1990).

A.4.2 Fish and shellfish

In 2 out of the 19 investigated fish and shellfish samples from Japan, Me_2-TBBPA was detected at levels of 0.8 and 4.6 μg/kg wet weight (Watanabe & Tatsukawa, 1990).

B. TETRABROMOBISPHENOL A DIBROMOPROPYLETHER

B.1 SUMMARY AND EVALUATION; CONCLUSIONS AND RECOMMENDATIONS

There is no data base on which to make an evaluation of tetrabromobisphenol A dibromopropylether, or to support its use commercially.

From the available data it can be concluded that the acute and short-term toxicities of tetrabromobisphenol A dibromopropylether are low. The substance was tested for mutagenicity and was a direct mutagen in *Salmonella typhimurium* strains TA100 and TA1535. However, the results of an unscheduled DNA synthesis assay and an *in vitro* Sister Chromatid Exchange test were negative.

This substance cannot be evaluated until adequate data become available on physical and chemical properties, production and use, environmental transport, distribution and transformation, environmental levels and human exposure, kinetics and metabolism in animals and humans, effects on laboratory mammals and humans, and effects on other organisms in the laboratory and field.

B.2 IDENTITY, PHYSICAL AND CHEMICAL PROPERTIES, AND ANALYTICAL METHODS

B.2.1 Identity

Chemical formula $C_{21}H_{20}Br_8O_2$

Chemical structure

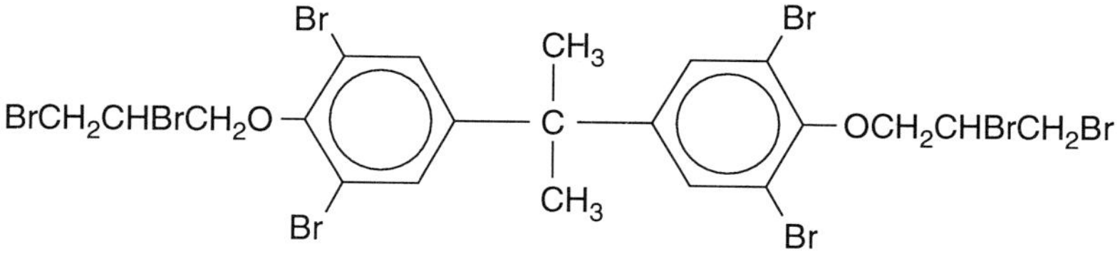

CAS registry number	21850-44-2
Synonyms	1,1′-(1-methylethylidene) bis (3,5-dibromo-4-(2,3-dibromopropoxy)-benzene; Bis(2,3-dibromopropoxy)-tetrabromobisphenol A; propane 2,2′-bis[3,5-dibromo-4-(2,3-dibromopropoxy)phenyl]; tetrabromobis phenol A dibromopropyl ether; 2,2′-bis[4-(2,3-dibromopropoxy)-3,5-dibromophenyl]-propane; bis(2,3-dibromo-propylether) of tetrabromobisphenol A; Dibromopropydian
Trade names	Bromcal 66.8; Fire guard 3100; PE-68

B.2.2 Physical and chemical properties

Tetrabromobisphenol A dibromopropylether is a crystalline or powdered white/off-white solid, with a slight odour. Decomposition takes place at temperatures > 270 °C. The bromine content is 68% (Arias, 1992). Other properties from Kopp (1990) are listed below.

Relative molecular mass	943.9

Melting point	90-100 °C (95 °C)
Specific gravity	0.7-0.9 g/cm^3
Solubility	1 g/litre water at 25 °C[a]

[a] This solubility seems to be too high.

B.3 SOURCES OF HUMAN AND ENVIRONMENTAL EXPOSURE

B.3.1 Uses

This substance is used as an additive flame retardant in polyolefins (Arias, 1992).

B.4 ENVIRONMENTAL TRANSPORT, DISTRIBUTION, AND TRANSFORMATION

Biodegradation tests have shown a negative response, and accumulation in carp was judged to be very small (Great Lakes Chemical Corporation, 1987, summary report).

No data are available on the following subjects:

- Environmental levels and human exposure
- Kinetics and metabolism in laboratory animals and humans
- Effects on humans
- Effects on other organisms in the laboratory and field

B.5 EFFECTS ON LABORATORY MAMMALS AND *IN VITRO* TEST SYSTEMS

No data are available on:

- Long-term exposure
- Skin and eye irritation; sensitization
- Reproductive toxicity, embryotoxicity, teratogenicity, and carcinogenicity

B.5.1 Single exposure

The acute LD_{50} for mice was > 20 g/kg when given in feed and observed for 14 days. The acute dermal LD_{50} for mice was > 20 g/kg when applied to closely clipped intact skin for 24 h and then observed for 14 days (Great Lakes Chemical Corporation, 1987, summary report).

B.5.2 Short-term exposures

Mice were administered levels of 200 or 2000 mg/kg per day in their diet for 90 days. At the end of the study, no deaths had occurred at either level. No abnormal symptoms were observed in the gross pathological examination (Great Lakes Chemical Corporation, 1987, summary report).

B.5.3 Mutagenicity and related end-points

B.5.3.1 Mutation

Three mutagenicity studies were carried out on *Salmonella typhimurium* strains TA1535, TA1537, TA1538, TA98, and TA100 with, and without, S9 fraction of livers of male Sprague-Dawley rats induced by Aroclor 1254. Eight dose levels of PE-68 (samples coded as 785-104A, 785-104B, and 785-104C) were used to test the mutagenicity, ranging from 1.00 μg to 10 000 μg/plate. DMSO was used as solvent. Samples 785-104A and 785-104C exhibited mutagenic activity with strains TA1535 and TA100 in the activation assay and with strains TA1535, TA100, and TA98 in the non-activation assay. Sample 785-104B exhibited mutagenic activity in the non-activation assay with TA1535 and TA100. These tests indicate that PE-68 is a direct-acting mutagen and that a rat liver S9 mix converts the test material to a less mutagenic form (Brusick, 1982).

B.5.3.2 Unscheduled DNA synthesis assay

PE-68 was tested in a rat (Sprague-Dawley) unscheduled DNA Synthesis Assay in duplicate doses of 10, 50, 100, 500, and 1000 μg/ml. The high dose was selected on the basis of the solubility of PE-68 in DMSO. No significant increase in the mean nuclear grain count was observed at any dose level compared with the solvent control. Positive medium and solvent controls confirmed the sensitivity of the system (Cavagnaro & Sernau, 1984).

B.5.3.3 In vitro sister chromatid exchange in Chinese hamster ovary cells

Chinese hamster ovary cells (CHO, K-1, number CCL61) were exposed to 5 concentrations of PE-68 in DMSO (5, 17, 50, 170, and 500 μg/ml) for 2 h in the presence, or absence, of metabolic activation followed by a 24-h expression period in comparison with solvent and positive controls. At dosing, it was noted that the culture medium became cloudy at 170 μg/ml and that the compound precipitated at 500 μg/ml. No statistically significant increases in the number of exchanges per chromosome or the number of exchanges per cell were seen at any of the levels tested, either with, or without, metabolic activation. PE-68 is considered to be negative in this system (Cavagnaro & Cortina, 1984).

C. TETRABROMOBISPHENOL A BIS(ALLYLETHER)

C.1. SUMMARY AND EVALUATION; CONCLUSIONS AND RECOMMENDATIONS

There is no data base on which to make an evaluation of tetrabromobisphenol A bis(allylether), or to support its use commercially.

From the available data it can be concluded that the acute oral and dermal toxicities of this compound are low. Skin and eye irritation studies on rabbits showed that the substance was a mild irritant for the eyes and skin.

This substance cannot be evaluated unless adequate data on physical and chemical properties, production and use, environmental transport, distribution, and transformation, environmental levels and human exposure, kinetics and metabolism in animals and humans, effects on laboratory mammals, humans, and *in vitro* test systems, and effects on other organisms in the laboratory and field, become available.

C.2 IDENTITY, PHYSICAL AND CHEMICAL PROPERTIES, AND ANALYTICAL METHODS

C.2.1 Identity

Chemical formula	$C_{21}H_{20}Br_4O_2$
Chemical structure	

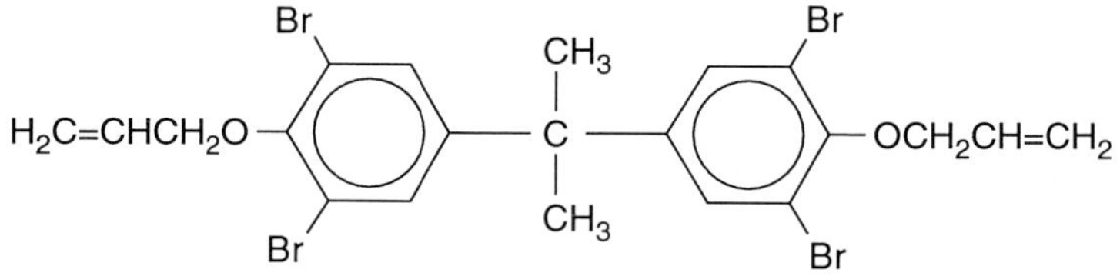

CAS registry number	25327-89-3
Trade names	BE-51

C.2.2 Physical and chemical properties

BE-51 is a crystalline white solid. The substance contains 51% bromine (Arias, 1992). By overheating, decomposition will take place with release of hydrogen bromide.

Relative molecular mass	655.9
Melting point	115-120 °C
Specific gravity	1.8
Solubility	< 1 g/litre water at 25 °C

C.2.3 Analytical methods

No data on this subject are available.

C.3 SOURCES OF HUMAN AND ENVIRONMENTAL EXPOSURE

C.3.1 Uses

It is used as a reactive flame retardant in polystyrene foams (Arias, 1992).

No data are available on the following subjects:

- Environmental transport, distribution, and transformation
- Environmental levels and human exposure
- Kinetics and metabolism in laboratory animals and humans
- Effects on humans
- Effects on other organisms in the laboratory and field

C.4 EFFECTS ON LABORATORY MAMMALS AND *IN VITRO* TEST SYSTEMS

No data are available on the following subjects:

- Short-term exposures
- Long-term exposure
- Reproductive toxicity, embryotoxicity, teratogenicity, and carcinogenicity

C.4.1 Single exposure

Sprague-Dawley rats (groups of 5 of each sex) (212-268 g) were administered (by gavage) a single dose of 5 g BE-51/kg body weight in corn oil. The observation time was 14 days. None of the rats died during the study and no signs of systemic toxicity were observed. The LD_{50} is > 5 g/kg body weight (Abbott et al., 1981).

A single dose of 2 g BE-51/kg body weight was applied to the clipped, abraded back skin of 5 male and 5 female young New Zealand albino rabbits (2.05-2.45 kg). The test site was covered with an occlusive wrap for 24 h. At the end of this period, the covering was removed and the residual test material wiped off. The observation time was 14 days. No animals died or exhibited signs of systemic toxicity, and body weight gain was normal. Slight to moderate erythema and oedema were observed. The skin reactions decreased in severity and area with time. No gross pathological findings were observed during necropsy. The dermal LD_{50} is > 2 g/kg body weight (Abbott et al., 1981).

C.4.2 Skin and eye irritation; sensitization

Young New Zealand albino rabbits (3 male and 3 female) (2.25-2.55 kg) were clipped on 4 sites on the back. Two test sites were abraded and 2 were left intact. A sample of 0.5 g BE-51, slightly moistened with physiological saline, was applied to each site and occluded for 24 h. At the end of this period, the covering was removed and the residual test material wiped off. The observation time was 4 days. Body weight gain was normal. No signs of systemic toxicity were observed and no deaths occurred. The primary skin irritation index was calculated to be 1.0 and

BE-51 was classified as mildly irritating using the Draize criteria for evaluation (Abbott et al., 1981).

The eyes of 4 male, and 5 female, young New Zealand albino rabbits (1.95-2.55 kg) were examined after the instillation of 0.1 g BE-51 in the conjunctival sac of one eye. The other eye was untreated and served as control. The treated eyes of one male and two female rabbits were flushed after 30 seconds with distilled water for 1 min. The eyes of the remaining rabbits were not flushed. Eye examinations for irritation were made at 24, 48, and 72 h, and, 4 and 7 days following application. No deaths occurred, body weight gain was normal, and no systemic toxicity symptoms were observed. Various degrees of swelling and redness were observed at the conjunctiva lasting for 4 days (not rinsed) and 48 h (rinsed). An irregular corneal surface (stippling) was observed in 6/9 of the rabbits. No signs of corneal damage were noted upon fluorescein examinations. The primary irritation index was determined to be 4.0 for unrinsed eyes and 1.33 for rinsed eyes. On the basis of these data, the substance is classified as mildly irritating for unrinsed eyes and minimally irritating for rinsed eyes (Abbott et al., 1981).

C.4.3 Mutagenicity and related end-points

A mutagenicity study was conducted with *Salmonella* and *Saccharomyces* indicator organisms, with, and without, metabolic activation with liver S9 fraction from Aroclor-induced rats. The dose levels of BE-5I ranged from 0.1 to 500 μg per plate. No mutagenic activity was found in this test (Brusick, 1977).

D. TETRABROMOBISPHENOL A BIS(2-HYDROXYETHYL ETHER)

D.1 SUMMARY AND EVALUATION; CONCLUSIONS AND RECOMMENDATIONS

The data base is inadequate for an evaluation of tetrabromobisphenol A bis(2-hydroxyethyl ether), or to support its use commercially.

From the available data, there is some indication that this substance may occur in the environment. The acute toxicity was low after oral and dermal administration in rats and rabbits, respectively. The acute inhalation toxicity (1-h exposure) in rats seemed to be moderate. A short-term toxicity study on rats showed no effects with 1000 mg/kg diet, but a significant increase in total bromine content in organs was observed. The substance was found not to irritate the skin and eyes of rabbits. The results of a mutagenicity study with five strains of *Salmonella typhimurium*, with, and without, metabolic activation, were negative.

The substance cannot be evaluated unless additional data on physical and chemical properties, production and use, environmental transport, distribution, and transformation, environmental levels and human exposure, kinetics and metabolism in animals and humans, effects on laboratory mammals, humans, and *in vitro* test systems, and effects on other organisms in the laboratory and field, become available. An *in vitro* cytogenetic study is also required.

D.2 IDENTITY, PHYSICAL AND CHEMICAL PROPERTIES, AND ANALYTICAL METHODS

D.2.1 Identity

Chemical formula $C_{19}H_{20}Br_4O_4$

Chemical structure

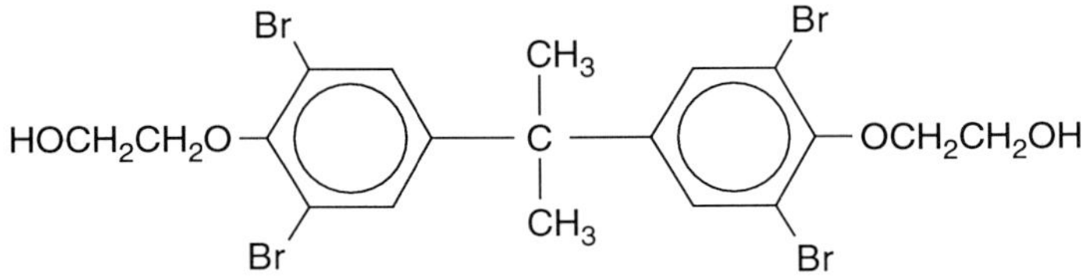

CAS registry number 4162-45-2

Trade names BA-50P; BA-50, Firegard 3600

D.2.2 Physical and chemical properties

The substance is a crystalline, white coloured, slightly chunky powder. It contains 51% bromine (Arias, 1992). BA-50P may release hydrogen bromide and/or bromine in fires fuelled by other products.

Melting point approximately 112 °C (115 °C)

Specific gravity approximately 1.80

D.3 SOURCES OF HUMAN AND ENVIRONMENTAL EXPOSURE

This substance is used as an additive flame retardant in engineering polymers and coatings (Arias, 1992).

D.4 ENVIRONMENTAL TRANSPORT, DISTRIBUTION, AND TRANSFORMATION

Carp were exposed to TBBPA bis(2-hydroxyethyl ether) at concentrations of 0.25 or 0.025 mg/litre, for 8 weeks. Bioaccumulation was 10.0-35.5 and 14.8-53.0, respectively (Chemicals Inspection and Testing Institute, 1992).

D.5 ENVIRONMENTAL LEVELS AND HUMAN EXPOSURE

D.5.1 Environmental levels

D.5.1.1 Air

No data are available.

D.5.1.2 Water

In 1986, an environmental survey was conducted concerning BA-50P in water at different locations in Japan. BA-50P was detected in 2 out of 30 samples at concentrations ranging from 20 to 40 μg/litre (limit of determination, 20 μg/litre) (Environment Agency Japan, 1989).

D.5.1.3 Soil

In 1986, an environmental survey was conducted concerning BA-50P in bottom sediment at different locations in Japan. BA-50P was not detected in 30 samples (limit of determination, 20 μg/kg dry weight) (Environment Agency Japan, 1989).

No data are available on the following subjects:

- Kinetics and metabolism in laboratory animals and humans
- Effects on humans
- Effects on other organisms in the laboratory and field

D.6 EFFECTS ON LABORATORY MAMMALS AND *IN VITRO* TEST SYSTEMS

No data are available on the following subjects:

- Long-term exposure
- Reproductive toxicity, embryotoxicity, teratogenicity, and carcinogenicity

D.6.1 Single exposure

Groups of five male Spartan rats (210-235 g) were administered (by gavage) 50, 500, or 5000 mg TBBPA bis(2-hydroxyethyl ether)/kg body weight. The compound was administered in corn oil at concentrations permitting a total dose of 10 ml/kg at all dose levels. The observation time was 14 days. None of the rats receiving 50 and 500 mg/kg died and they exhibited normal body weight gain. In the 5000 mg/kg group, 2 out of 5 rats died during the 14 days, and the three remaining rats showed decreased body weight gain. The acute oral LD_{50} was > 5.0 g/kg body weight (Goldenthal & Dean, 1974).

TBBPA bis(2-hydroxyethyl ether) was applied to the closely clipped intact skin of two male and two female New Zealand albino rabbits (2339-2937 g) each at dose levels of 200 or 2000 mg/kg body weight. The application site was occluded for 24 h. The bandages were removed and the backs were washed with tepid tap water. The observation time was 14 days. No rabbits died during the study. At 200 and 2000 mg/kg, 3 out of 4 rabbits gained weight while 1 out of 4 lost weight in each group. The acute dermal LD_{50} for rabbits is > 2 g/kg body weight (Goldenthal & Dean, 1974).

Two groups of 5 male and 5 female Spartan rats (210-236 g) were exposed for 1 h to a dynamic atmosphere containing TBBPA bis(2-hydroxyethyl ether) dust at calculated concentrations of 2 or 12.5 mg/litre of air. The observation time was 14 days. Nine out of 10 rats at the 2 mg/litre dose survived and appeared normal during the first 8 days; one rat showed slight dyspnoea on days 4 and 5. From days 9-14, one or two rats exhibited ocular discharge or drying of the corneal surface for a few days. The rats showed normal body weight gain. All rats exposed to 12.5 mg/litre survived and exhibited normal growth. During exposure, rats showed a slight increase in motor activity for the first 10 min of exposure and eye squint. At 24 h and up to day 8 of the 14-day observation period, all rats appeared normal, with the exception of one rat that exhibited marked and slight dyspnoea on days 4 and 5, respectively. From day 9 onwards, one or two rats exhibited drying of the corneal surface accompanied by eye squint. The LC_{50} for inhalation of TBBPA bis(2-hydroxyethyl ether) dust is > 12.5 mg/litre (Goldenthal & Dean, 1974).

D.6.2 Short-term exposures

Charles River CD rats (males 296-392 g; females 205-257 g) were fed dietary levels of 0, 100, or 1000 mg TBBPA bis(2-hydroxyethyl ether)/kg for 28 days. There were 10 male and 10 female rats in each group. Feed consumption and body weight gain were recorded. At the end of the study, all rats were killed. Besides gross pathological examination, the liver, kidneys, and thyroid were examined microscopically. Liver and fat tissues were pooled according to sex and dose groups for bromine determination. None of the rats died, and no changes were noted in the behaviour or appearance of any of the rats during the study. Feed consumption and body weight gain were normal. A slight increase in the bromine contents (5.0-7.3 mg/kg) of the liver was seen, but not in the fat tissue of the rats receiving 100 mg/kg. At the 1000 mg/kg level, a definite increase in total bromine content was seen in both the liver (18.3-48.8 mg/kg) and fat (4.4-22.1 mg/kg) tissue. No compound-related changes in organ weights, gross pathological lesions, or histopathological changes were observed in the liver, kidneys, or thyroid, in any of the rats (Goldenthal & Geil, 1974).

D.6.3 Skin and eye irritation; sensitization

Three male and three female New Zealand albino rabbits (2682-2998 g) had 500 mg TBBPA bis(2-hydroxyethyl ether) applied to the closely clipped intact skin (3 rabbits) or to the closely clipped abraded skin (other three rabbits). The application site was occluded for 24 h, after which the rabbits' skin was washed and examined for skin irritation. The examinations were repeated at 72 h. At 24 and 72 h, 1/3 rabbits in the intact group exhibited very slight erythema or very slight oedema. No erythema or oedema was observed in the rabbits with the abraded skin. The calculated primary irritation score was 0.2 and indicated that the substance was not a primary skin irritant (Goldenthal & Dean, 1974).

Single applications of 100 mg TBBPA bis(2-hydroxyethyl ether) were made into the conjunctival sac of one eye of three male and three female New Zealand albino rabbits (2443-2670 g). Examinations were done at 24, 48, and 72 h and at 7 days. At 72 days, fluorescein and UVR were used to detect corneal damage. No corneal damage, iridal irritation, or conjunctival discharge was noted. Very slight to slight redness and very slight chemosis of the conjunctivae were noted in some of the animals at 24, 48, and 72 h, with decreasing frequency until all rabbits appeared normal at 7 days. These results indicate that the substance is not an eye irritant (Goldenthal, & Dean, 1974).

D.6.4 Mutagenicity and related end-points

TBBPA bis(2-hydroxyethyl ether) was examined for mutagenic activity in the microbial assays using *Salmonella typhimurium* TA1535, TA1537, TA1538, TA98, and TA100 in the presence, or absence, of liver microsomal enzyme preparations from Aroclor 1254-induced rats. The concentrations tested were 0, 0.5, 1, 10, 100, 500, and 1000 μg/plate, in comparison with positive control substances. The results indicated that the substance was not mutagenic under these test conditions (Jagannath & Brusick, 1979).

E. TETRABROMOBISPHENOL A BROMINATED EPOXY OLIGOMER

E.1 SUMMARY AND EVALUATION; CONCLUSIONS AND RECOMMENDATIONS

The data base is inadequate for an evaluation of tetrabromobisphenol A brominated epoxy oligomer, and to support its use commercially.

Some, but insufficient, data on the physical and chemical properties and production and use of tetrabromobisphenol A brominated epoxy oligomer are available. The quantities of PBDD and PBDF produced when resins containing these epoxy oligomers were pyrolised were much lower than those produced when TBBPA was pyrolysed.

These substances cannot be evaluated unless adequate data on their physical and chemical properties, production and use, environmental transport, distribution, and transformation, environmental levels and human exposure, kinetics and metabolism in laboratory animals and humans, effects on laboratory mammals, humans, and *in vitro* test systems, and effects on other organisms in the laboratory and field, become available.

As the use of these compounds seems to be increasing, at least in Japan, it is essential that further studies are performed.

E.2 IDENTITY, PHYSICAL AND CHEMICAL PROPERTIES, AND ANALYTICAL METHODS

E.2.1 Identity

There are two chemically different types of brominated epoxy oligomers. One has two epoxy groups at the end of the molecule, which is quite similar to epoxy resins used for printed circuit boards (EP-type). The other has no reactive groups. This is TBBPA epoxy end-capped with tribromophenol (EC-type) (Satoh & Sugie, 1993).

Chemical structure

EP type (Epoxy terminated):

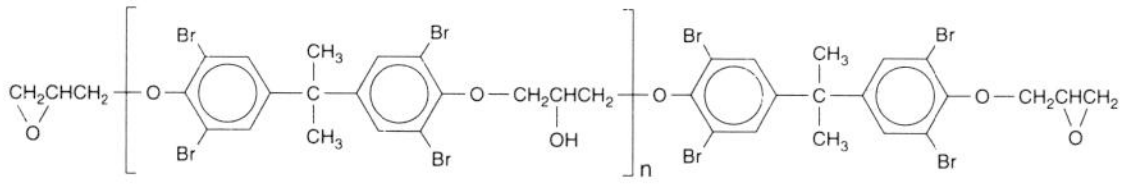

n = 0, 1, 2, 3 ...

EC type (tribromophenol end-capped):

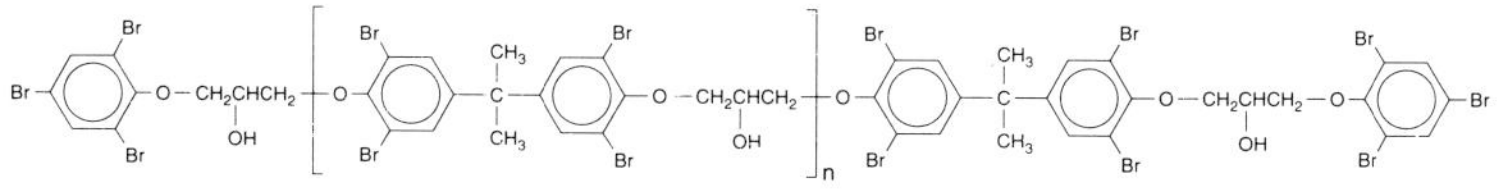

n = 0, 1, 2, 3 ...

E.2.2 Physical and chemical properties

	EP Type	EC Type
Relative molecular mass	1.300-40.000	1.400-3.000
Appearance	light yellow powder	light yellow powder
Specific gravity	1.8	1.9
Bromine contents (%)	50-52	59-55
Softening point (°C)	103-> 200	99-140

From: Satoh & Sugie (1993).

E.2.3 Analytical methods

No data are available.

E.3 SOURCES OF HUMAN AND ENVIRONMENTAL EXPOSURE

E.3.1 Natural occurrence

Brominated epoxy oligomers have not been reported to occur naturally.

E.3.2 Anthropogenic sources

E.3.2.1 Production levels and processes

Brominated epoxy oligomer flame retardants were first introduced on the Japanese market in 1987, with a demand of approximately 3000 tonnes in 1991. Demand requirements still show a rapid growth in Japan as well as in the USA.

The products are especially characterized by a higher melt flow rate without blooming, and a better light stability than existing flame retardants, such as polybrominated diphenylethers and others (Satoh & Sugie, 1993).

E.3.2.2 Uses

Brominated epoxy oligomers are reactive flame retardants. They have been applied in housings for business machinery and electrical/electronics parts by injection moulding from flame retardant compounds based upon HIPS, ABS, ABS/PC, or PBT alloys, PBT, and thermosetting resins. A new application is for use in large-size TV sets, moulded from HIPS.

The concentrations of the flame retardant in ABS are 21% of the EP-type and 19% of the EC-type. Brominated epoxy oligomers are used in combination with 5% of Sb_2O_3 (Satoh & Sugie, 1993).

E.4 ENVIRONMENTAL TRANSPORT, DISTRIBUTION, AND TRANSFORMATION

E.4.1 Pyrolysis of polymers containing brominated epoxy oligomers

ABS containing the EP-type and EC-type brominated epoxy oligomers and Sb_2O_3 were heated at 600 °C. Gas and ash were collected and analysed for the presence of PBDF and PBDD. In this study, TBBPA was also tested for comparison. TBBPA produced 0.9 μg 2,3,7,8-PBDD/kg including PeBDD and HxBDD, and 22 μg 2,3,7,8-PBDF/kg. The EP-type gave < 0.5 μg/kg (sum of TeBDD, PeBDD, and HxBDD) and the EC-type < 4 μg/kg. The values for the sum of TeBDF, PeBDF, and HxBDF were 0.5 and < 4 μg/kg, respectively (Satoh & Sugie, 1993).

No data are available on the following subjects:

- Environmental levels and human exposure
- Kinetics and metabolism in laboratory animals and humans
- Effects on laboratory mammals and *in vitro* test systems
- Effects on humans
- Effects on other organisms in the laboratory and field

F. TETRABROMOBISPHENOL A CARBONATE OLIGOMERS

F.1 SUMMARY AND EVALUATION; CONCLUSIONS AND RECOMMENDATIONS

There is no data base on which to make an evaluation of tetrabromobisphenol A carbonate oligomer, or to support its use commercially.

The results of mutagenicity studies with five strains of *Salmonella typhimurium*, with, and without, metabolic activation, were negative for both substances.

These substances cannot be evaluated unless adequate data on physical and chemical properties, production and use, environmental transport, distribution, and transformation, environmental levels and human exposure, kinetics and metabolism in animals and humans, effects on laboratory mammals, humans, and *in vitro* test systems, and effects on other organisms in the laboratory and field, become available. *In vitro* cytogenetic studies are also required.

F.2 IDENTITY, PHYSICAL AND CHEMICAL PROPERTIES, ANALYTICAL METHODS

F.2.1 Identity of BC-52

Chemical formula $(C_7H_5O_2)(C_{16}H_{10}Br_4O_3)_x$
x = 3-5

Chemical structure

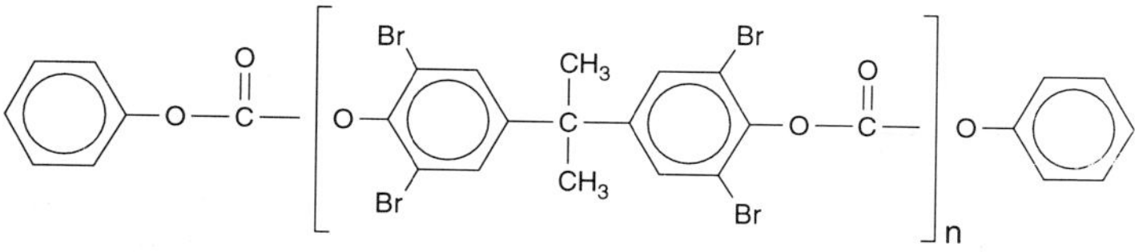

n = 3 – 5

CAS registry number 94334-64-2

F.2.1.1 Physical and chemical properties

BC-52 is a white powder. Its bromine content is 55%. It may release hydrogen bromide and/or bromine in fires fuelled by other products.

In BC-52, 6 ng/kg TeBDD was found. No PBDF and no 2,3,7,8-substituted isomers were detected (limits of detection ranged from 1 to 400 ng/kg for DiBDF/DiBDD to OBDF/OBDD) (Brenner & Knies, 1993).

Melting point 210-230 °C

Solubility in water < 0.1% in water at 25 °C

From: Kopp (1990); Arias (1992).

F.2.2 Identity of BC-58

Chemical formula $(C_7H_2Br_3O_3)(C_{16}H_{10}Br_4O_3)_n(C_6H_2Br_3)$

Chemical structure

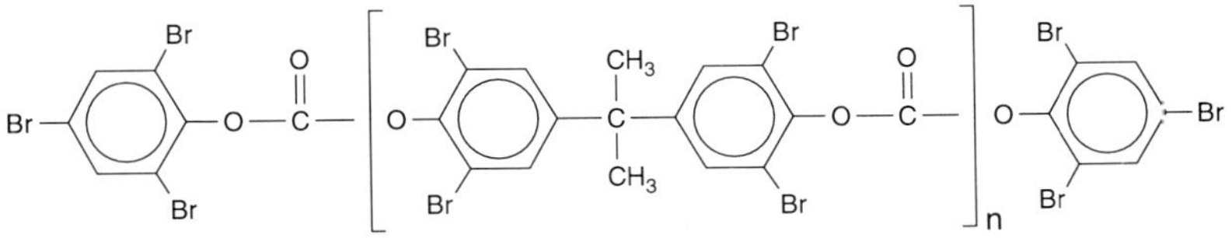

n = 3 – 5

CAS registry number 71342-77-3

F.2.2.1 Physical and chemical properties

BC-58 is a white powder. It may release hydrogen bromide and/or bromine in fires fuelled by other products. The following properties are from Kopp (1990).

Melting point	230-260 °C
Specific gravity	2.2
Solubility in water	negligible

F.3 SOURCES OF HUMAN AND ENVIRONMENTAL EXPOSURE

F.3.1 Uses

These oligomers are used as an additive flame retardant in engineering thermoplastics and ABS (Arias, 1992).

F.4 ENVIRONMENTAL TRANSPORT, DISTRIBUTION, AND TRANSFORMATION

F.4.1 Transport and distribution

No data are available on environmental transport and distribution.

F.4.2 Transformation

F.4.2.1 Pyrolysis

In general, the pyrolysis of tetrabromobisphenol A oligomer in combination with antimony trioxide at temperatures of 200, 250, or 600 °C, over 30 min, resulted in only low levels (up to 15 mg/kg) of PBDF. No PBDD was found (Fresenius Institute, 1990).

F.4.2.2 Monitoring of PBDF/PBDD during extrusion blending and injection moulding

Thies et al. (1990) reported investigations into the processing of a polymer containing polybutylene-terephthalate, 10% TBBPA - oligocarbonate, and 5% antimony trioxide. PBDD/PBDF levels were determined in the polymer and the condensate after a 20-min treatment at 240 °C in the BIS-apparatus. Mono- and di-BDF were detected at 13 and 10 μg/kg, respectively, but levels of other PBDD/PBDF were less than 2 μg/kg.

Brenner & Knies (1993) analysed PBDF and PBDD: a) during the extrusion of PBT blended with BC-52 powder (BC-52; ca. 11%/Sb_2O_3 ca. 4%), b) during the extrusion of BC-52 PBT-batch (ca. 50% BC-52), and, c) during injection moulding of the produced PBT/glass-fibre/BC-52/Sb_2O_3 resin.

Air monitoring was performed at the workplace, and the exhaust air was measured at the extruder and injection heads in the granulator exhaust line.

The results of the extruder experiment with BC-52 powder (PBDF concentrations in ng/m^3) are shown in Table A.

Table A. PBDF concentrations in ng/m^3[a]

	DiBDF	TrBDF	TeBDF	PeBDF	HxBDF	HpBDF
Workplace	0.34	0.11	0.05	0.07	0.05	n.d.[b]
Extruder head	0.42	0.48	0.24	0.04	0.15	n.d.[b]
Granulator	0.23	0.29	0.17	0.02	n.d.[b]	n.d.[b]

[a] From: Brenner & Knies (1993).
[b] n.d. = not detected (below limit of detection).

DiBDD, TrBDD, and TeBDD were found in concentrations of 0.94, 0.07, and 0.08 ng/m^3, respectively, at the extruder head, and, 0.02 ng/m^3 DiBDD in the granulator. The levels of all other PBDDs were below the limit of detection.

The results of the extruder experiment with PBT/BC-52 (use of BC-52 batch) are shown in Table B.

Table B. PBDF concentrations in ng/m^3[a]

	DiBDF	TrBDF	TeBDF	PeBDF	HxBDF	HpBDF
Workplace	0.14	0.35	0.08	0.05	0.02	n.d.[b]
Extruder head	0.16	0.31	0.06	0.333	0.01	n.d.[b]
Granulator	0.2	0.42	0.08	0.07	0.24	n.d.[b]

[a] From: Brenner & Knies (1993).
[b] n.d. = not detected (below limit of detection).

DiBDD concentrations of 0.001, 0.007, and 0.003 ng/m^3 were detected at monitoring points at the workplace, extruder head, and granulator.

The results of the injection moulding experiment performed with PBT/glass fibre/BC-52/Sb_2O_3 granulate are shown in Table C.

Table C. PBDF concentrations in ng/m^3[a]

	DiBDF	TrBDF	TeBDF	PeBDF	HxBDF	HpBDF
Workplace	n.d.[b]	n.d.[b]	0.029	0.187	0.262	n.d.[b]
Injection head	0.004	0.012	0.014	0.013	0.039	n.d.[b]
Storage	0.004	n.d.[b]	0.002	n.d.[b]	n.d.[b]	n.d.[b]

[a] From: Brenner & Knies (1993).
[b] n.d. = not detected (below limit of detection).

PBDD were also found including: DiBDD at < 1 pg/m^3 and OBDD at < 232 pg/m^3 (limits of detection). No 2,3,7,8-TeBDD could be detected (limit of detection 0.001-0.058 ng/m^3) (Brenner & Knies, 1993).

F.4.2.3 PBDD/PBDF levels in polymer samples using BC52-powder, BC52-batch, and the moulded test articles produced from these

Concentrations of PBDF in extruder granulate (PBT-granulate) using BC52 powder, and the moulded test articles produced from that, are listed in Table D.

Table D. PBDF concentrations in $\mu g/kg$[a]

	DiBDF	TrBDF	TeBDF	PeBDF	HxBDF	HpBDF
PBT granulate (three samples)	n.d.[b]	n.d.[b]	n.d.[b]	n.d.[b]	0.4-0.8	0.6-3.5
Test article A	0.07	0.2	0.2	n.d.[b]	2.2	3.8
Test article B	0.29	0.31	0.17	0.06	1.5	1.9

[a] From: Brenner & Knies (1993).
[b] n.d. = not detected.

Finally, PBDF was determined in the BC-52/PBT batch and the product produced (PBT-granulate). DiBDF was found in the two batch samples at concentrations of 1.0 and 1.4 μg/kg and in the granulate at 0.6 μg/kg. No other PBDF were found (Brenner & Knies, 1993).

F.5 ENVIRONMENTAL LEVELS AND HUMAN EXPOSURE

See section F.4.2.2 for monitoring at the workplace.

F.6 EFFECTS ON LABORATORY MAMMALS AND *IN VITRO* TEST SYSTEMS

F.6.1 Single exposure

The oral LD_{50} for BC-52 and BC-58 in the rat is > 5 g/kg, and, the dermal LD_{50} for the rabbit > 2.0 g/kg body weight (Kopp, 1990).

F.6.2 Skin and eye irritation; sensitization

BC-52 and BC-58 are not primary skin or eye irritants (Kopp, 1990).

F.6.3 Mutagenicity and related end-points

Both substances were tested in 5 strains of *Salmonella typhimurium* at doses ranging from 100 to 10 000 μg/plate, in the presence, and absence, of metabolic activation. Both gave negative results (Great Lakes 1983a,b).

No data are available on the following subjects:

- Short-term exposures
- Long-term exposure
- Reproductive toxicity, embryotoxicity, teratogenicity, and carcinogenicity
- Kinetics and metabolism in laboratory animals and humans
- Effects on humans
- Effects on other organisms in the laboratory and field

REFERENCES

Abbott L, Altringer L, Kingery AF, & Mayhew DA (1981) Acute dermal LD_{50} toxicity study in albino rabbits, acute eye irritation study in albino rabbits, acute oral LD_{50} toxicity study in albino rats, primary skin irritation study in albino rabbits with BE-51. Cincinnati, Ohio, Wil Research Laboratories, Inc. (Unpublished report No. WIL-81225 to Great Lakes Chemical Corporation, West Lafayette, submitted to WHO by the Brominated Flame Retardant Industry Panel).

Allard A-S, Remberger M, & Neilson AH (1987) Bacterial O-methylation of halogen-substituted phenols. Appl Environ Microbiol, **53**(4): 839-845.

Anon (1974) Markets for organo-bromo flame retardants depend on legislation. Eur Chem News, December 6: 48.

Arias P (1992) Brominated diphenyloxides as flame retardant: Bromine based chemicals (Unpublished report to the Organisation for Economic Cooperation and Development, Paris).

Bayer (1990) Chemical dossier on tetrabromobisphenol A. Leverkusen, Germany, Bayer AG (Unpublished report).

Brady UE (1979) Pharmacokinetic study of tetrabromobisphenol A (BP-4A) in rats. Report from Athens, Georgie, University of Georgia (Report to Velsicol Chemical Corporation, Chicago, submitted to WHO by the Brominated Flame Retardant Industry Panel).

Brenner KS & Knies H (1993) Workplace monitoring of PBDFs and PBDDs during extrusion production and injection molding of a polybutyleneterephthalte (PBTP)/glass fiber/tetrabromobisphenol A carbonate oligomer (BC52*)/Sb_2O_3-resin; Part II. Chemosphere, **26**(11): 1953-1963.

Breteler RJ (1989) The subchronic toxicity of sediment-sorbed tetrabromobisphenol A (*Chironomus tentans*) under flow-through conditions. Wareham, Massachusetts, Springborn Laboratories, Inc. (Report No. 90-08-3067 submitted to WHO by the Brominated Flame Retardant Industry Panel).

Brusick D (1976) Mutagenicity evaluation of compound 279-117-2 (Final report). Report of Kensington, Maryland, Litton Bionetics, Inc. (Report to Great Lakes Chemical Corporation, West Lafayette, submitted to WHO by the Brominated Flame Retardant Industry Panel).

Brusick D (1977) Mutagenicity evaluation of BE-51. Kensington, Maryland, Litton Bionetics, Inc. (Report to Great Lake Chemical Corporation, West Lafayette, submitted to WHO by the Brominated Flame Retardant Industry Panel).

Brusick D (1982) Mutagenicity evaluation of 785-104A, 785-104B and 785-104C in the Ames *Salmonella*/microsome plate test (Final report). Kensington, Maryland, Litton Bionetics, Inc. (Combined Reports Nos. LBI 7655, 7656, and 7657 to Great Lakes Chemical Corporation, West Lafayette, submitted to WHO by the Brominated Flame Retardant Industry Panel).

Calmbacher CW (1978a) The acute toxicity of FMBP4A (tetrabromobisphenol A) to the bluegill sunfish, *Lepomis macrochirus* Rafinesque. Tarrytown, New York, Union Carbide Corporation, Environmental Services (Report to Velsicol Chemical Corporation, Chicago, submitted to WHO by the Brominated Flame Retardant Industry Panel).

Calmbacher CW (1978b) The acute toxicity of FMBP4A (tetrabromobisphenol A) to the rainbow trout, *Salmo gairdneri* Richardson. Tarrytown, New York, Union Carbide Corporation, Environmental Services (Report to Velsicol Chemical Corporation, Chicago, submitted to WHO by the Brominated Flame Retardant Industry Panel).

Cavagnaro J & Cortina TA (1984) *In vitro* sister chromatid exchange in Chinese hamster ovary cells with GLCC 785-104C (Final report). Vienna, Virginia, Hazleton Biotechnologies Corporation (Report to Great Lakes Chemical Corporation, West Lafayette, submitted to WHO by the Brominated Flame Retardant Industry Panel).

Cavagnaro J & Sernau RC (1984) Unscheduled DNA synthesis rat hepatocyte assay, GLCC 785-104C. (Final report). Vienna, Virginia, Hazleton Biotechnologies Corporation (Report to Great Lakes Chemical Corporation, West Lafayette, submitted to WHO by the Brominated Flame Retardant Industry Panel).

Chemicals Inspection & Testing Institute (1992) Biodegradation and bioaccumulation data of existing chemicals based on the CSCL, Japan. Tokyo, Japan Chemical Industry Ecology, Toxicology and Information Center, pp 4-14.

Clausen E, Lahaniatis ES, Bahadir M, & Bieniek D (1987) [Determination of brominated dibenzofurans formed during the thermolysis of polymer with decabromodiphenyl ether as flame retardant.] Fresenius Z Anal Chem, **327**: 297-300 (in German).

Craig DK, Mitchum RK, Bauer MR, Yancey MF, Peters AC, & Joiner RL (1989) Determination of polybrominated dibenzo-p-dioxins and polybrominated dibenzofurans in CYCOLAC plastic resins and the fumes evolved during normal thermal processing (Final report). Columbus, Ohio, Batelle (Report No. 823-R-4289 to General Electric Company, Mt. Vernon, submitted to WHO by the Brominated Flame Retardant Industry Panel).

Dean WP, Jessup DC, Epstein WL, & Powell D (1978a) Tetrabromobisphenol A. Modified Draize multiple insult test in humans. Mattawan, Michigan, International Research and Development Corporation (Report to Velsicol Chemical Corporation, Chicago, submitted to WHO by the Brominated Flame Retardant Industry Panel).

Dean WP, Jessup DC, Thompson G, Romig G, & Powell D (1978b) Tetrabromobisphenol A. Dermal sensitization study in the albino guinea-pig. Mattawan, Michigan, International Research and Development Corporation (Report to Velsicol Chemical Corporation, Chicago, submitted to WHO by the Brominated Flame Retardant Industry Panel).

Dumler R, Thoma H, Lenoir D, & Hutzinger O (1989) PBDF and PBDD from the combustion of bromine containing flame retarded polymers: A survey. Chemosphere, **19**(12): 2023-2031.

Environment Agency Japan (1989) Chemicals in the environment. Report on environmental survey and wildlife monitoring of chemicals in F.Y. 1986 and 1987. Tokyo, Environment Agency Japan, Department of Environmental Health, Office of Health Studies.

Environment Agency Japan (1991) Chemicals in the environment. Report on environmental survey and wildlife monitoring of chemicals in F.Y. 1988 and 1989. Tokyo, Environment Agency Japan, Department of Environmental Health, Office of Health Studies.

Ethyl Corporation (1992a) Material safety data sheet for emergencies. Baton Rouge, Louisiana, Ethyl Corporation, Chemicals Group (Report submitted to WHO by the Brominated Flame Retardant Industry Panel).

Ethyl Corporation (1992b) Saytex RB-100. Baton Rouge, Louisiana, Ethyl Corporation (Report submitted to WHO by the Brominated Flame Retardant Industry Panel).

Fackler PH (1989a) Determination of the biodegradability of tetrabromobisphenol A in soil under aerobic conditions (Final report). Wareham, Massachusetts, Springborn Life Sciences, Inc. (Report No. 88-11-2848 submitted to WHO by the Brominated Flame Retardant Industry Panel).

Fackler PH (1989b) Bioconcentration and elimination of ^{14}C-residues by Eastern oysters (*Crassostrea virginica*) exposed to tetrabromobisphenol A. Wareham, Massachusetts, Springborn Life Sciences, Inc. (Report No. 89-1-2918 submitted to WHO by the Brominated Flame Retardant Industry Panel).

Fackler PH (1989c) Bioconcentration and elimination of ^{14}C-residues by fathead minnows (*Pimephales promelas*) exposed to tetrabromobisphenol A. Wareham, Massachusetts, Springborn Life Sciences, Inc. (Report No. 89-3-2952 submitted to WHO by the Brominated Flame Retardant Industry Panel).

Fackler PH (1989d) Determination of the biodegradability of tetrabromobisphenol A in soil under anaerobic conditions. Wareham, Massachusetts, Springborn Life Sciences, Inc. (Report No. 88-11-2849 submitted to WHO by the Brominated Flame Retardant Industry Panel).

Fackler PH (1989e) (Tetrabromobisphenol A) - Determination of the biodegradability in a sediment/soil microbial system. Wareham, Massachusetts, Springborn Life Sciences, Inc. (Report No. SLI 89-8-3070 submitted to WHO by the Brominated Flame Retardant Industry Panel).

Fresenius Institute (1990) Summary results on pyrolysis of different types of ABS/Batelle report on contents and vapour-emission of PBDD resp. PBDF's. Taunusstein-Neuhof, Germany, Fresenius Institute (Report submitted to WHO by the Brominated Flame Retardant Industry Panel).

Giddings JM (1988) Toxicity of tetrabromobisphenol A to the freshwater green alga *Selenastrum capricornutum*. Wareham, Massachusetts, Springborn Life Sciences, Inc. (Report No. 88-10-2828 submitted to WHO by the Brominated Flame Retardant Industry Panel).

Goldenthal EI & Dean WP (1974) Bis(2-hydroxyethyl ether) of tetrabromobisphenol A. Acute toxicity studies in rats and rabbits. Mattawan, Michigan, International Research and Development Corporation (Report to Great Lakes Chemical Corporation, West Lafayette, submitted to WHO by the Brominated Flame Retardant Industry Panel).

Goldenthal EI & Geil RG (1972) Tetrabromobisphenol A. Twenty-eight day toxicity study in rats. Mattawan, Michigan, International Research and Development Corporation (Report to Great Lakes Chemical Corporation, West Lafayette, submitted to WHO by the Brominated Flame Retardant Industry Panel).

Goldenthal EI & Geil RG (1974) Bis(2-hydroxyethyl ether) of tetrabromobisphenol A; Decabromodiphenyl ether, 250-139-2; decabromodiphenyl ether, 143-78-6. Twenty-eight day toxicity study in rats. Mattawan, Michigan, International Research and Development Corporation (Report to Great Lakes Chemical Corporation, West Lafayette, submitted to WHO by the Brominated Flame Retardant Industry Panel).

Goldenthal EI, Geil RG, & Dean WP (1975) Tetrabromobisphenol A (micronized). Fourteen-days inhalation toxicity study in rats. Mattawan, Michigan, International Research and Development Corporation (Report to Great Lakes Chemical Corporation, West Lafayette, submitted to WHO by the Brominated Flame Retardant Industry Panel).

Goldenthal EI, Jessup DC, & Rodwell DE (1978) Tetrabromobisphenol A (FMBP-4A). Pilot teratology study in rats. Mattawan, Michigan, International Research and Development Corporation (Report to Velsicol Chemical Corporation, Chicago, submitted to WHO by the Brominated Flame Retardant Industry Panel).

Goldenthal EI, Jessup DC, Geil RG, Dean WP, & Ruecker FA (1979) BP-4A. Three-week dermal toxicity study in rabbits. Mattawan, Michigan, International Research and Development Corporation (Report to Velsicol Chemical Corporation, Chicago, submitted to WHO by the Brominated Flame Retardant Industry Panel).

Goodman LR, Cripe GM, Moody PH, & Halsell DG (1988) Acute toxicity of malathion, tetrabromobisphenol A and tributyltin chloride to Mysids, (*Mysidopses bahia*) of three ages. Bull Environ Contam Toxicol, **41**: 746-753.

Great Lakes Chemical Corporation (1983a) Summaries of toxicity data. BC-52. West Lafayette, Indiana, Great Lakes Chemical Corporation (Unpublished report submitted to WHO by the Brominated Flame Retardant Industry Panel).

Great Lakes Chemical Corporation (1983b) Summaries of toxicity data. BC-58. West Lafayette, Indiana, Great Lakes Chemical Corporation (Unpublished report submitted to WHO by the Brominated Flame Retardant Industry Panel).

Great Lakes Chemical Corporation (1986) Summaries of toxicity data. Tetrabromobisphenol A. West Lafayette, Indiana, Great Lakes Chemical Corporation (Unpublished report submitted to WHO by the Brominated Flame Retardant Industry Panel).

Great Lakes Chemical Corporation (1987) Summaries of toxicity data. PE-68, Bis(2,3-dibromopropyl ether) of tetrabromobisphenol A. West Lafayette, Indiana, Great Lakes Chemical Corporation (Unpublished report submitted to WHO by the Brominated Flame Retardant Industry Panel).

Gustafsson K & Wallen M (1988) Status report on tetrabromobisphenol A (CAS no. 79-94-7). Clearing house Sweden. Solna, Sweden, National Chemicals Inspectorate (Unpublished report).

Hardy ML (1994) Summary; TBBPA toxicological studies sponsored by Ethyl Corporation. Baton Rouge, Louisiana, Ethyl Corporation (Report submitted to WHO by the Brominated Flame Retardant Industry Panel).

Inouye B, Katayama Y, Ishida T, Ogata M, & Utsumi K (1979) Effects of aromatic bromine compounds on the function of biological membranes. Toxicol Appl Pharmacol, **48**: 467-477.

Jagannath DR & Brusick DJ (1979) Mutagenicity evaluation of No. 341-150 in the Ames *Salmonella*/microsome plate test (Final report). Kensington, Maryland, Litton Bionetics, Inc. (Report No. LBI No. 20988 to Great Lakes Chemical Corporation, West Lafayette, submitted to WHO by the Brominated Flame Retardant Industry Panel).

Kawamura K, Takeuchi T, & Kobayashi S (1986) Inhibition of respiratory activities of *Giardia lamblia* by halogenated bisphenols. Jpn J Parasitol, **35**(3): 261-263.

Kopp A (1990) [Documentation on brominated flame retardants.] Bonn, Federal Ministry for Environment, Nature Conservation and Reactor Safety (Prepared for the European Economic Community, Brussels) (in German).

Lahaniatis ES, Bergheim W, & Bieniek D (1991) Formation of 2,3,7,8-tetrabromo-dibenzodioxin and -furan by thermolysis of polymers containing brominated flame retardants. Toxicol Environ Chem, **31/32**: 521-526.

Lorenz W & Bahadir M (1993) Recycling of Flame Retardants containing printed circuits: a study of the possible formation of polyhalogenated dibenzodioxins/-furans. Chemosphere, **26**(12): 2221-2229.

Luijk R & Govers HAJ (1992) The formation of polybrominated dibenzo-p-dioxins (PBDDs) and dibenzofurans (PBDFs) during pyrolysis of polymer blends containing brominated flame retardants. Chemosphere, **25**(3): 361-374.

Meyer H, Neupert M, Pump W, & Willenberg B (1993) [Flame retardants determine reusability.] Kunststoffe, **83**(4): 253-257 (in German).

Morrissey AE (1978) The acute toxicity of FMBP4A (tetrabromobisphenol A) to the water flea, *Daphnia magna* Straus. Tarrytown, New York, Union Carbide Corporation, Environmental Services (Report to Velsicol Chemical Corporation, Chicago, submitted to WHO by the Brominated Flame Retardant Industry Panel).

Mortelmans K, Haworth S, Lawlor T, Speck W, Tainer B, & Zeiger E (1986) *Salmonella* mutagenicity test: II. Results from the testing of 270 chemicals. Environ Mutagen, **8**(Suppl 7): 1-119.

Neupert M & Pump W (1992) Experiences from a large scale warehouse fire with bromine-containing polybutylene terephthalate forming almost no polybrominated dibenzodioxins and dibenzofurans. Leverkusen, Germany, Bayer AG (Unpublished document).

Noda T (1985) Safety evaluation of chemicals for use in household products (VII): Teratological studies on tetrabromobisphenol-A in rats. Annu Rep Osaka City Inst Public Health Environ Sci, **48**: 106-121.

Nye DE (1978) The bioaccumulation of tetrabromobisphenol, in the bluegill sunfish. Santa Clara, California, Stoner Laboratories, Inc. (Report to Velsicol Chemical Corporation, Chicago, submitted to WHO by the Brominated Flame Retardant Industry Panel).

OECD (1993) Brominated flame retardants: Draft status report. Paris, Organisation for Economic Cooperation and Development.

Quast JF, Humiston CG, & Schwetz BA (1975) Results of a 90-day toxicological study in rats given tetrabromobisphenol A in the diet. Midland, Michigan, Dow Chemical (Unpublished report No. HET 17.5-36-(3), submitted to WHO by the Brominated Flame Retardant Industry Panel).

Ranken P (1993) The issue of tetrabromobisphenol A (TBBPA) and brominated dioxins and furans. Baton Rouge, Louisiana, Ethyl Corporation, Health and Environmental Department (Report submitted to WHO by the Brominated Flame Retardant Industry Panel).

Satoh Y & Sugie K (1993) Brominated epoxy oligomer flame retardants - as one of a new generation of FRs. Tokyo, Japan, Dainippon Ink and Chemicals, Inc., Petrochemicals Division (Unpublished paper presented at the OECD Workshop on Brominated Flame Retardants, Neuchâtel, Switzerland, February 1993).

Sellström U, Jansson B, Jonsson P, Nylund K, Odsjö T, & Olsson M (1990) Anthropogenic brominated aromatics in the Swedish environment. In: Short papers - Dioxin 1990, EPRI Seminar, Bayreuth, Germany, pp 357-360.

Sellström U, Jansson B, & Zakrisson S (1994) [Analysis of tetrabromobisphenol-A in product and environmental testing.] Solna, Sweden, State Nature Conservation Agency (Unpublished document) (in Swedish).

Steinberg CEW, Sturm A, Kelbel J, Kyu Lee S, Hertkorn N, Freitag D, & Kettrup AA (1992) Changes of acute toxicity of organic chemicals to *Daphnia magna* in the present of dissolved humic material. Acta Hydrochim Hydrobiol, **20**(6): 326-332.

Sterner W (1967) Acute oral toxicity of tetrabromo-bis-phenol A to rats; acute inhalation toxicity study of tetrabromo-bis-phenol A and acute eye irritation study on rabbits of tetrabromo-bis-phenol A. St. Louis, Missouri, International Bio-Research, Inc. (Report to Great Lakes Chemical Corporation, West Lafayette, submitted to WHO by the Brominated Flame Retardant Industry Panel).

Surprenant DC (1988) Acute toxicity of tetrabromobisphenol A to fathead minnow (*Pimephales promelas*) under flow-through conditions. Wareham, Massachusetts, Springborn Life Sciences, Inc. (Report No. SLS 88-10-2834 submitted to WHO by the Brominated Flame Retardant Industry Panel).

Surprenant DC (1989a) The toxicity of tetrabromobisphenol A (TBBPA) to fathead minnow (*Pimephales promelas*) embryos and larvae. Wareham, Massachusetts, Springborn Life Sciences, Inc. (Report No. 89-2-2937 submitted to WHO by the Brominated Flame Retardant Industry Panel).

Surprenant DC (1989b) The chronic toxicity of tetrabromobisphenol A (TBBPA) to *Daphnia magna* under flow-through conditions. Wareham, Massachusetts, Springborn Life Sciences, Inc. (Report No. 89-01-2925 submitted to WHO by the Brominated Flame Retardant Industry Panel).

Surprenant DC (1989c) Acute toxicity of tetrabromobisphenol A to Eastern oysters (*Crassostrea virginica*) under flow-through conditions. Wareham, Massachusetts, Springborn Life Sciences, Inc. (Report No. 89-1-2898 submitted to WHO by the Brominated Flame Retardant Industry Panel).

Tatsukawa R & Watanabe I (1990) [Environmental problems from brominated organic flame retardants.] Kogai Taisaku, **26**(7): 658-668 (in Japanese)

The Chemical Daily (1990-1994) [Special report on flame retardants.] Chem Daily (in Japanese).

Thies J, Neupert M, & Pump W (1990) Tetrabromobisphenol A (TBBA), its derivatives and their flame retarded (FR) polymers - content of polybrominated dibenzo-p-dioxins (PBDD) and dibenzofurans (PBDF) - PBDD/F formation under processing and smouldering (worst case) conditions. Chemosphere, **20**(10-12): 1921-1928.

Thoma H, Rist S, Hauschultz G, & Hutzinger O (1986a) Polybrominated dibenzodioxins (PBrDD) and Dibenzofurans (PBrDF) in some flame retardant preparations, Chemosphere, **15**(9-12), 2111-2113.

Thoma H, Rist S, Hauschulz G, & Hutzinger O (1986b) Polybrominated dibenzodioxins and -furans from the pyrolysis of some flame retardants. Chemosphere, **15**(5): 649-652.

Tobe M, Kurokawa Y, Nakaji Y, Yoshimoto H, Takagi A, Aida Y, Monma J, Naito K, & Saito M (1986) [Subchronic toxicity study of tetrabromobisphenol-A: Report to the Ministry of Health and Welfare] (in Japanese).

Tondeur Y, Mazac C, Freiberg M, Ranken P, Hass R, & McAllister D (1990) Analytical procedures for the determination of polybrominated dibenzo-p-dioxins and dibenzofurans in tetrabromobisphenol A and 2,4,6-tribromophenol. Chemosphere, **20**: 373-376.

Ullmann T (1985) In: Gerhartz W, Yamamoto YS, Campbell FT, Pfefferkorn R, & Rounsaville JF ed. Ullmann's encyclopedia of industrial chemistry, 5th revis ed. Weinheim, Germany, VCH Verlagsgesellschaft, vol A4.

Walsh GE, Yoder MJ, McLaughlin LL, & Lores EM (1987) Responses of marine unicellular algae to brominated organic compounds in six growth media. Ecotoxicol Environ Saf, **14**: 215-222.

Watanabe I & Tatsukawa R (1990) Anthropogenic brominated aromatics in the Japanese environment. In: Freij L ed. Proceedings of Workshop on Brominated Aromatic Flame Retardants, Skokloster, 24-26 October 1989. Solna, Sweden, National Chemicals Inspectorate, pp 63-71.

Watanabe I, Kashimoto T, & Tatsukawa R (1983a) Identification of the flame retardant tetrabromobisphenol A in the river sediment and the mussel collected in Osaka. Bull Environ Contam Toxicol, **31**: 48-52.

Watanabe I, Kashimoto T, & Tatsukawa R (1983b) The flame retardant tetrabromobisphenol A and its metabolite found in river and marine sediments in Japan. Chemosphere, **12**(11-12): 1533-1539.

Zweidinger RA, Cooper SD, Erickson MD, Michael LC, & Pellizzari ED (1979) Sampling and analysis for semi volatile brominated organics in ambient air. In: Schuetzle D ed. Monitoring toxic substances. Washington, DC, American Chemical Society, pp 217-231 (ACS Symposium Series 94).

RESUME ET EVALUATION; CONCLUSIONS ET RECOMMANDATIONS RELATIVES AU TETRABROMOBISPHENOL A (TBBPA)

1. Résumé et évaluation

1.1 Propriétés physiques et chimiques

Le TBBPA se présente sous la forme d'une poudre cristalline blanche (incolore) contenant 59% de brome. Son point de fusion est d'environ 180 °C et son point d'ébullition de 316 °C. Sa tension de vapeur est très inférieure à 1 mmHg à 20 °C. Le TBBPA est peu soluble dans l'eau mais très soluble dans le méthanol et dans l'acétone. Son coefficient de partage *n*-octanol/eau (log P_{ow}) est égal à 4,5.

1.2 Production et usages

De tous les retardateurs de flamme bromés, c'est le TBBPA du commerce qui est le plus produit dans le monde. La demande de TBBPA et de ses dérivés représente plus de 60 000 tonnes par an. On l'utilise comme retardateur de flammes réactif (usage principal) ou également pour cet usage, comme simple additif, dans les polymères tels que l'ABS, les résines époxy, les polycarbonates, le polystyrène choc, les résines phénoliques, les adhésifs, etc.

1.3 Transport, distribution et transformation dans l'environnement

Compte tenu de son coefficient de partage et de sa faible solubilité dans l'eau, le TBBPA présent dans l'environnement devrait être sujet à une importante sorption sur les sédiments et les matières organiques présents dans le sol. L'étude de son accumulation dans les invertébrés et les vertébrés aquatiques montre que le facteur de bioconcentration varie de 20 à 3200. Sa demi-vie est de moins de 1 jour chez les poissons et de moins de 5 jour chez les huîtres. Au cours de la dépuration, la majeure partie du TBBPA accumulé (et de ses métabolites) seront éliminés dans les 3 à 7 jours.

Des études de biodégradation ont montré que le TBBPA était partiellement décomposé, dans des conditions aérobies ou anaérobies, dans le sol ainsi que dans les sédiments des cours d'eau et dans l'eau. Selon la nature, la température, le degré d'humidité et la composition du sol, on a constaté que le TBBPA y subsistait encore, à hauteur d'environ 40 à 90%, au bout de 56 à 64 jours. Dans les conditions qu'entraîne le traitement des effluents, on n'a pas constaté de biodégradation d'après la mesure de la DBO à 2 semaines.

En étudiant la décomposition, par pyrolyse en laboratoire, de polymères contenant du TBBPA, en l'absence ou en présence de Sb_2O_3, à différentes températures ainsi qu'en présence d'oxygène, etc., on a constaté qu'il pouvait se former des dibenzofuranes polybromés (BBDF) et, dans une moindre mesure, des dibenzodioxines polybromées (BPDD). L'analyse de polymères additionnés de TBBPA, et placés dans les conditions simulant un traitement thermique, n'a pas permis de mettre en évidence de 2,3,7,8-PBDD ou PBDF. On a seulement mis en évidence des PBDF mono- ou dibromo-substitués à des concentrations allant jusqu'à 100 μg/kg de résine. Des analyses effectuées sur l'air des lieux de travail n'ont pas permis non plus de déceler la présence de PBDD ou de PBDF substitués en position 2,3,7,8 (limite de détection égale 0,1 ng/m^3).

L'analyse de polymères recyclés contenant du TBBPA a mis en évidence moins de 5 μg de PBDF/PBDD totaux par kg et les congénères substitués en position 2,3,7,8 ne s'y trouvaient qu'à des concentrations inférieures à 0,2 μg/kg.

Lors de l'incendie d'un entrepôt au cours duquel une grande quantité de téréphtalate de polybutylène (PBT) contenant du TBBPA, a brûlé, on n'a décelé dans les résidus brûlés de PBT ainsi que dans des échantillons de cendres et de matières fondues, que de faibles quantités de tétra-, penta-, hexa-BDF ou BDD substitués en position 2,3,7,8 (soit moins de 5 μg/kg).

1.4 Concentrations dans l'environnement et exposition humaine

Au Japon et en Suède, on a décelé la présence de TBBPA dans certains sédiments et au Japon encore, dans des poissons (dans deux échantillons sur 229 à proximité d'une zone industrielle), en quantités de l'ordre du mg/kg. Dans des moules et des sédiments, on a pu mettre en évidence la présence du dérivé diméthoxy du TBBPA. En général, on n'a pas trouvé de TBBPA dans l'eau.

1.5 Cinétique et métabolisme chez les animaux de laboratoire et l'homme

Chez le rat, la TBPPA est faiblement absorbé et au niveau des voies digestives. Une fois résorbé, le composé initial et ses métabolites se répartissent dans la plupart des organes. Chez le rat, on a observé, quel que soit le tissu, une demi-vie inférieure à 2,5 jours.

1.6 Effets sur les mammifères de laboratoire et les systèmes d'épreuve in vitro

Le TBBPA ne présente qu'une faible toxicité aiguë par voie orale pour les animaux de laboratoire. Ainsi des études ont montré que la DL_{50} par voie orale pour le rat était > 5 g/kg de poids corporel et qu'elle était de 10 g/kg de poids corporel pour la souris. Chez la lapin, la DL_{50} par voie percutanée s'est révélée > 2 g/kg de poids corporel. Par inhalation, le CL_{50} pour la souris, le rat et le cobaye s'est révélée > 0,5 mg/litre. Une seule application cutanée de TBBPA, à des concentrations allant jusqu'à 3,16 g/kg de poids corporel, à des lapins et à des cobayes, n'a pas provoqué d'effets locaux ou généralisés. Le TBBPA ne s'est pas révélé irritant pour la peau ni pour les yeux chez le lapin. Les quelques études portant sur des cobayes n'ont pas permis de mettre en évidence des réactions de sensibilisation. On a également recherché les possibilités d'induction d'une chloracné par le TBBPA sur l'oreille du lapin. Aucune réaction de ce genre n'a été observée. Une étude de toxicité percutanée de 3 semaines au cours de laquelle on a badigeonné la peau rasée et abrasée de lapins avec du TBBPA à des concentrations allant jusqu'à 2500 mg/kg de poids corporel, n'a mis en évidence qu'un léger érythème cutané. Aucune autre altération imputable à ce composé n'a été observée.

Des rats ont été exposés à des concentrations allant jusqu'à 18 mg/litre de TBBPA micronisé (18 000 mg/m^3) pendant 2 semaines, 4 heures par jour, 5 jours par semaine. Aucun effet n'a été observé, qu'il s'agisse du poids corporel, des résultats des analyses sanguines et des analyses d'urine, des constantes chimiques sériques ou de l'histologie.

Des doses de TBBPA allant jusqu'à 1000 mg/kg de nourriture ont été administrées pendant 28 jours à des rats par voie orale sans produire le moindre effet indésirable. Il n'y avait aucune différence entre les groupes témoins et les groupes soumis aux doses élevées (1000 mg/kg) en ce qui concerne la teneur du foie en brome.

Lors d'une étude de toxicité de 90 jours au cours de laquelle on a fait ingérer à des rats des doses de TBBPA allant jusqu'à

100 mg/kg de poids corporel, on n'a pas constaté d'effet indésirable sur le poids corporel ou le poids des organes, les paramètres hématologiques, les constantes chimiques, les résultats de l'analyse d'urine ainsi que ceux de l'examen histologique et macroscopique.

Des souris à qui l'on a fait ingérer pendant 90 jours une dose de 4900 mg de TBBPA/kg de nourriture (soit environ 700 mg/kg de poids corporel par jour) n'ont pas subi d'effet indésirable; en revanche une dose de 15 600 mg/kg de nourriture (soit environ 2200 mg/kg de poids corporel par jour) a entraîné une réduction du poids corporel, une augmentation du poids de la rate, une diminution de la concentration des hématies ainsi que de celle des protéines et des triglycérides sériques.

Deux études de tératogénicité ont été effectuées sur des rats; l'une au cours de laquelle on a administré par gavage des concentrations allant jusqu'à 10 g/kg de poids corporel du 6ème au 15ème jour de la gestation et une deuxième au cours de laquelle des doses allant jusqu'à 2,5 g/kg de poids corporel ont été administrées du jour zéro au jour 19 de la gestation. Dans la première étude, trois animaux sur cinq ayant reçu 10 g/kg de TBBPA sont morts, en revanche aucun signe de toxicité n'a été relevé chez les animaux qui en avaient reçu 3 g/kg. Aucun effet tératogène n'a été observé. Quant à la deuxième étude, elle n'a pas révélé d'anomalies.

Pour étudier le pouvoir mutagène éventuel du TBBPA, on a utilisé dans diverses études, les souches de *Salmonella typhimurium* TA1535, TA1537, TA1538, TA98 et TA100, l'activation métabolique étant obtenue au moyen d'un mélange enzymatique S9 provenant de rats et de hamsters de Syrie traités par l'Aroclor; les résultats de ces études ont été négatifs. Les concentrations utilisées allaient jusqu'à 10 000 μg/boîte. Deux épreuves effectuées sur *Saccharomyces cerevisiae* avec ou sans préparation enzymatique microsomienne provenant de rats traités par l'Aroclor, ont également donné des résultats négatifs.

Il n'a pas été fait état d'études de cancérogénicité ou de toxicité à long terme.

1.7 Effets sur l'homme

Le TBBP n'a pas produit d'irritation ou de sensibilisation cutanée chez 54 volontaires humains.

On ne dispose d'aucune étude épidémiologique ni d'autres types de données sur les effets de ce produit chez l'homme.

1.8 *Effets sur les autres êtres vivants au laboratoire et dans leur milieu naturel*

Le TBBPA ne s'est pas révélé très toxique pour les algues marines. Les valeurs de la CE_{50} tirées de 28 études à court terme se situaient dans les limites de 0,1 à 1,0 mg/litre, alors que pour les algues d'eau douce, on n'observait aucune inhibition de croissance, même à la concentration de 9,6 mg/litre.

Pour *Daphnia magna*, la CL_{50} aiguë à 48 heures serait de 0,96 mg/litre; à la dose de 0,32 mg/litre, 5% des animaux sont morts. Lors d'une étude de 21 jours, toutefois, on a constaté que la CE_{50} correspondant à la survie et la croissance de *Daphnia magna* était > 0,98 mg/litre. En s'appuyant sur les effets du TBBPA sur la reproduction des daphnies, constatés au cours de cette même étude, on a obtenu une concentration de substances toxiques maximale acceptable se situant entre 0,30 et 0,98 mg/litre. Pour les mysidés (respectivement âgés de < 1, 5 et 10 jours), les valeurs de CL_{50} à 96 heures étaient respectivement égales à 0,86, 1,1 et 1,2 mg/litre.

Chez une espèce d'huître, on a calculé que la CE_{50} à 96 heures (réduction de la formation de la coquille par dépôt calcaire) était de 0,098 mg/litre avec une concentration sans effets observables de 0,0062 mg/litre.

La CL_{50} à 96 heures du TBBPA pour trois espèces de poissons: *Lepomis macrochirus*, la truite arc-en-ciel et *Pimephales promelas* était respectivement égale à 0,51, 0,40 et 0,54 mg/litre. Les concentrations sans effets observables pour les trois espèces de poissons étaient respectivement égales à 0,10, 0,18 et 0,26 mg/litre. Des embryons et des larves de Pimephales promelas ont été exposés 35 jours à du TBBPA et on a constaté que la concentration maximale acceptable de substance toxique se situait entre 0,16 et 0,31 mg/litre, à en juger d'après les effets délétères du TBBPA sur la survie de ces embryons et de ces larves.

En ce qui concerne un invertébré des sédiments, *Chironomous tentans*, on a obtenu comme valeurs de la concentration sans effet à 14 jours, respectivement 0,039, 0,045 et 0,046 mg de TBBPA/litre d'eau dans des sédiments de faible, moyenne et forte teneur en carbone organique.

La plupart des études portant sur des organismes aquatiques ont été réalisées à des pH voisins du pKa_2. Dans des eaux acides, le comportement du TBBPA pourrait être différent.

2. Conclusions

2.1 Population générale

On utilise largement le TBBPA incorporé à des polymères soit sous forme d'additif, soit sous forme réactive, comme retardateur de flamme. La population générale peut entrer en contact avec cette substance par l'intermédiaire d'objets dans la composition desquels entrent ces polymères, sans que cela entraîne une absorption notable de TBBPA. En outre, le toxicité du TBBPA est très faible, qu'elle soit aiguë ou chronique. Ce composé est difficilement absorbé au niveau des voies digestives. On peut donc en conclure que le risque résultant, pour la population générale, d'une exposition au TBBPA, est à considérer comme insignifiant.

2.2 Exposition professionnelle

L'exposition professionnelle au TBBPA se produit essentiellement par contact avec des particules lors de l'incorporation de ce produit comme additif, ou lors de l'emballage. Le dépoussiérage des locaux grâce à une bonne ventilation ou toute autre technique, permet de réduire le risque pour les ouvriers. S'il n'est pas possible d'éliminer convenablement les poussières, les ouvriers devront protéger leurs voies respiratoires.

2.3 Environnement

Le TBBPA que l'on retrouve dans l'environnement est essentiellement présent dans le sol et les sédiments. La valeur relativement élevée de son facteur de bioconcentration semble être compensée par une excrétion rapide et ce composé n'est normalement pas présent dans les échantillons biologiques prélevés dans l'environnement.

Dans l'environnement, il peut y avoir méthylation des groupements phénoliques du TBBPA, conduisant à du Me_2-TBBPA qui est plus lipophile. On a également retrouvé ce composé dans des sédiments, dans du poisson, des mollusques et des crustacés.

2.4 Produits de décomposition

Des traces de PBDD et de PBDF peuvent se retrouver dans le TBBPA comme impuretés; toutefois on n'a pas pu mettre en évidence la présence de congénères substitués en position 2,3,7,8. Par pyrolyse au laboratoire, le TBBPA donne naissance à des PBDF et à des PBDD.

Quelques études ont montré que, lorsqu'on procède à la transformation et au recyclage de polymères contenant du TBBPA sous forme d'additif retardateur de flammes, il ne se forme que des traces de ces dernières substances.

3. Recommandations

3.1 Recommandations générales

- Les personnes qui sont employées à la fabrication du TBBPA et de produits qui en contiennent doivent être protégées contre toute exposition à cette substance par la mise en oeuvre de moyens techniques appropriés, la surveillance de l'exposition professionnelle et des mesures d'hygiène convenables.

- Un traitement approprié des effluents et des émissions provenant d'industries utilisant ce composé ou les produits qui en contiennent devrait permettre de réduire au minimum les cas d'exposition dans l'environnement.

- Le rejet des déchets industriels et des produits de consommation devrait être contrôlé afin de réduire au minimum la contamination de l'environnement par ce composé et ses produits de décomposition.

- Pour incinérer des produits contenant du TBBPA, on utilisera des appareils bien conçus fonctionnant toujours dans des conditions optimales.

3.2 Etudes à effectuer

- Continuer à surveiller la présence de TBBPA, Me_2-TBBPA, PBDF et PBDD dans l'environnement par analyse d'échantillons et en cas de résultats positifs, procéder également à une surveillance chez l'homme.

- Surveiller l'exposition professionnelle à des particules respirables de TBBPA; si les résultats obtenus sur les lieux de travail l'exigent, procéder à une étude d'inhalation à court terme sur des rats.

- Etudier la formation de PBDF et de PBDD à partir de produits traités par du TBBPA lors d'opérations d'incinération, incendies accidentels ou dans des conditions qui les simulent.

- Procéder à des études à long terme sur la destinée des polymères contenant du TBBPA (soit sous forme d'additif, soit sous forme réactive), notamment dans les décharges contrôlées.

- Etudier la transformation dans l'environnement, du TBBPA en son dérivé diméthylé.

- Poursuivre l'étude des possibilités de recyclage des polymères contenant du TBBPA, en accordant une attention particulière aux produits de décomposition.

- Vu l'absence de données, il est nécessaire de procéder à une épreuve *in vitro* supplémentaire avec le TBBPA, à la recherche de lésions cytogénétiques éventuelles. En cas de résultats positifs, il sera nécessaire de procéder à d'autres études *in vivo*. Si ces études donnent à leur tour des résultats positifs, il faudra procéder à des épreuves à court et à long terme complémentaires.

- Vu l'absence de données, il est nécessaire d'étudier la toxicité du produit sur la reproduction du rat.

ETHER DIMETHYLIQUE DU TETRABROMODISPHENOL A

Il n'existe aucune base de données sur laquelle s'appuyer pour procéder à l'évaluation de l'éther diméthylique du tétrabromobisphénol A ni pour en justifier l'usage commercial.

Il n'est pas possible d'évaluer l'éther diméthylique du tétrabromobisphénol A tant qu'on ne dispose pas de données suffisantes sur ses propriétés physiques et chimiques, sa production et son usage, son transport, sa distribution, sa transformation et sa concentration dans l'environnement et l'exposition humaine auxquels ils peuvent donner lieu, sa cinétique et son métabolisme chez l'animal et l'homme, ses effets sur les mammifères de laboratoire, sur l'homme ainsi que sur les systèmes d'épreuve *in vitro*; enfin, son action sur les autres êtres vivants, tant au laboratoire que dans leur milieu naturel.

ETHER DIBROMOPROPYLIQUE DU TETRABROMODISPHENOL A

Il n'existe aucune base de données sur laquelle s'appuyer pour évaluer l'éther dibromopropylique du tétrabromobisphénol A ni pour en justifier l'usage commercial.

On peut déduire des données disponibles que la toxicité aiguë et à court terme de l'éther dibromopropylique du tétrabromobisphénol A est faible. Cette substance a fait l'objet d'une épreuve de mutagénicité et l'on a constaté qu'elle se comportait comme un mutagène direct vis-à-vis des souches de *Salmonella typhimurium* TA100 et TA1535. Toutefois, les résultats d'une épreuve de synthèse non programmée de l'ADN et la recherche d'échanges entre chromatides soeurs *in vitro*, ont été négatifs.

Le produit ne pourra pas être évalué tant qu'on ne disposera pas de données suffisantes sur ses propriétés physiques et chimiques, sa production et son usage, son transport, sa distribution, sa transformation et sa concentration dans l'environnement et l'exposition humaine auxquels ils peuvent donner lieu, sa cinétique et son métabolisme chez l'animal et l'homme, ses effets sur les animaux de laboratoire et l'homme et enfin son action sur les autres êtres vivants au laboratoire et dans leur milieu naturel.

ETHER DIALLYLIQUE DE TETRABROMODISPHENOL A

Il n'existe pas de base de données sur laquelle s'appuyer pour évaluer l'éther diallylique de tétrabromobisphénol A ni pour en justifier l'usage commercial.

D'après les données disponibles, on peut conclure que la toxicité aiguë par voie orale et la toxicité percutanée de ce composé sont faibles. Des études d'irritation cutanée et oculaire chez le lapin ont montré que le composé était légèrement irritant à ce niveau.

Ce produit ne pourra pas être évalué tant qu'on ne disposera pas de données suffisantes sur ses propriétés physiques et chimiques, sa production et son usage, son transport, sa distribution, sa transformation et sa concentration dans l'environnement ainsi que l'exposition humaine à laquelle ils peuvent donner lieu, sa cinétique et son métabolisme chez l'animal et l'homme, ses effets sur les animaux de laboratoire et sur l'homme ainsi que sur les systèmes d'épreuve *in vitro* et enfin, son action sur les autres êtres vivants au laboratoire et dans leur milieu naturel.

ETHER BIS(2-HYDROXYETHYLIQUE) DE TETRABROMODISPHENOL A

La base de données est insuffisante pour permettre d'évaluer l'éther bis(2-hydroxyéthylique) de tétrabromobisphénol A ou en justifier l'usage commercial.

D'après les données disponibles il semblerait que cette substance puisse être présente dans l'environnement. Après administration par voie orale et percutanée respectivement à des rats et à des lapins, on a constaté que sa toxicité aiguë était faible. Il semble également que sa toxicité aiguë par voie respiratoire (1 heure d'exposition) soit modérée chez le rat. Une étude de toxicité à court terme sur des rats n'a pas permis de mettre d'effet en évidence à la dose de 1000 mg/kg de nourriture, toutefois on a observé une augmentation sensible de la teneur en brome total des organes. Le produit n'est pas irritant au niveau cutané ou oculaire chez le lapin. Les résultats d'une étude de mutagénicité sur 5 souches de *Salmonella typhimurium*, avec ou sans activation métabolique, ont été négatifs.

Ce produit ne pourra pas être évalué tant qu'on ne disposera pas de données complémentaires sur ses propriétés physiques et chimiques, sa production et son usage, son transport, sa distribution, sa transformation et sa concentration dans l'environnement ainsi que l'exposition humaine à laquelle ils peuvent donner lieu, sa cinétique et son métabolisme chez l'animal et l'homme, ses effets sur les animaux de laboratoire et l'homme ainsi que sur les systèmes d'épreuve *in vitro* et enfin, son action sur les autres êtres vivants au laboratoire et dans leur milieu naturel. Il est également nécessaire de procéder à une étude cytogénétique *in vitro*.

OLIGOMERE EPOXYDIQUE BROME DU TETRABROMODISPHENOL A

La base de données est insuffisante pour permettre d'évaluer l'oligomère époxydique bromé du tétrabromobisphénol A ou pour justifier son usage commercial.

On dispose de quelques données - encore qu'insuffisantes - sur les propriétés physiques et chimiques et sur la production et l'usage de l'oligomère époxydique bromé du tétrabromobisphénol A. On a constaté que les quantités de PBDD et de PBDF produites lors de la pyrolyse de résines contenant ces oligomères, sont beaucoup plus faibles que celles obtenues par pyrolyse du TBBPA.

Il ne sera pas possible d'évaluer ces produits tant qu'on ne disposera pas de données suffisantes sur leurs propriétés physiques et chimiques, leur production et leur usage, leur transport, leur distribution, leur transformation et leur concentration dans l'environnement et l'exposition humaine à laquelle ils peuvent donner lieu, leur cinétique et leur métabolisme chez l'animal et l'homme, leurs effets sur les animaux de laboratoire, l'homme et les systèmes d'épreuve *in vitro* et enfin, leur action sur les autres êtres vivants au laboratoire et dans leur milieu naturel.

Comme il semble que l'on utilise de plus en plus ces composés, tout au moins au Japon, il est essentiel qu'ils soient étudiés plus avant.

OLIGOMERES DE CARBONATE DE TETRABROMODISPHENOL A

Il n'existe pas de base de données sur laquelle s'appuyer pour d'évaluer les oligomères de carbonate de tétrabromobisphénol A ni pour en justifier l'usage commercial.

Les résultats d'études de mutagénicité portant sur cinq souches de *Salmonella typhimurium*, avec ou sans activation métabolique, se sont révélés négatifs pour ces substances.

Il ne sera pas possible d'évaluer ces composés tant qu'on ne disposera pas de données suffisantes sur leurs propriétés physiques et chimiques, leur production et leur usage, leur transport, leur distribution, leur transformation et leur concentration dans l'environnement et l'exposition humaine à laquelle ils peuvent donner lieu, leur cinétique et leur métabolisme chez l'animal et l'homme, leurs effets sur les animaux de laboratoire, l'homme et les systèmes d'épreuve *in vitro* et enfin, leur action sur les autres êtres vivants au laboratoire et dans leur milieu naturel. Il sera également nécessaire de procéder à des études cytogénétiques.

Comme il semble que l'on utilise de plus en plus ces composés, tout au moins au Japon, il est essentiel qu'ils soient étudiés plus avant.

RESUMEN Y EVALUACION; CONCLUSIONES Y RECOMENDACIONES SOBRE EL TETRABROMOBISFENOL A (TBBFA)

1. Resumen y evaluación

1.1 Propiedades físicas y químicas

El TBBFA es un polvo blanco (incoloro), cristalino, que contiene un 59% de bromo. Su punto de fusión es de aproximadamente 180 °C, y su punto de ebullición, 316 °C. Su presión de vapor es muy inferior a 1 mmHg a 20 °C. El TBBFA es poco soluble en agua, pero es muy soluble en metanol y en acetona. El coeficiente de reparto *n*-octanol/agua (log P_{ow}) es de 4,5.

1.2 Producción y utilización

El TBBFA es el pirorretardante bromado que se produce en mayor cantidad en el mundo. La demanda de TBBFA y de sus derivados supera las 60 000 toneladas al año. El TBBFA se utiliza como reactivo (utilización principal) o como aditivo pirorretardante en polímeros tales como las resinas ABS, epoxi y policarbonatos, poliestireno de alta resistencia al impacto, resinas fenólicas, adhesivos y otras sustancias.

1.3 Transporte, distribución y transformación en el medio ambiente

Debido a su coeficiente de reparto y a su escasa solubilidad en agua, se prevé que el TBBFA en el medio ambiente sea en gran medida objeto de sorción en el sedimento y en la materia orgánica del suelo.

Los estudios sobre acumulación realizados en animales acuáticos invertebrados y vertebrados indican factores de bioconcentración que oscilan entre 20 y 3200. La semivida en peces es de menos de un día y en ostras es de menos de cinco días. Durante la depuración, la mayor parte del TBBFA acumulado (y sus metabolitos) se eliminan a los 3-7 días.

Los estudios sobre biodegradación mostraron que el TBBFA se degrada parcialmente en condiciones aerobias y anaerobias en el suelo y en el sedimento y el agua de los ríos. Según el tipo de suelo, su temperatura, humedad y composición, aproximadamente el 40-90% del TBBFA permanece en el suelo a los 56-64 días. En condiciones de tratamiento de aguas residuales, no se detectó biodegradación a las dos semanas cuando la biodedegradación se midió como demanda biológica de oxígeno.

Los estudios sobre pirólisis en laboratorio mostraron que los polímeros que contienen TBBFA pueden, con o sin presencia de Sb_2O_3, a diferentes temperaturas, en presencia de oxígeno, etc., formar dibenzofuranos polibromados (DFPB) y, en menor medida, dibenzodioxinas polibromadas (DDPB). Se forman principalmente DFPB y DDPB débilmente bromados. Cuando se analizaron polímeros formulados con TBBFA que se habían expuesto a condiciones de procesamiento térmico estimulante, no se detectaron 2,3,7,8-DDPB/DFPB. Sólo se detectaron DFPB con uno o dos átomos de bromo de sustitución en niveles de hasta 100 μg/kg en la resina. Las investigaciones realizadas en el medio laboral no mostraron DDPB/DFPB con sustitución en las posiciones 2,3,7,8 (límite de detección: 0,1 ng/m^3).

En los polímeros reciclados que contienen TBBFA se detectaron en total menos de 5 μg de DFPB/DDPB por kg y los congéneres con sustitución en las posiciones 2,3,7,8 se encontraron solamente en niveles inferiores a 0,2 μg/kg.

Tras un incendio ocurrido en un depósito en el que se quemó una gran cantidad de tereftalato de polibutileno (TPB) que contenía TBBFA, se detectaron solamente niveles bajos de dibenzofuranos tetrabromados, pentabromados y hexabromados y de dibenzodioxinas tetrabromadas, pentabromadas y hexabromadas con sustitución en las posiciones 2,3,7,8 (menos de 5 μg/kg) en el TPB quemado y en las muestras de cenizas/escoria.

1.4 Niveles ambientales y exposición humana

Se detectó la presencia de TBBFA en algunos sedimentos en el Japón y en Suecia y en peces (en dos muestras extraídas cerca de una zona industrial, de un total de 229 muestras) en niveles del orden de μg/kg en el Japón. Pudo identificarse el derivado dimetoxi del TBBFA en mejillones y en el sedimento. En general, no se detectó la presencia de TBBFA en el agua.

1.5 Cinética y metabolismo en animales de laboratorio y en seres humanos

En las ratas, la absorción del TBBFA a través del tracto gastrointestinal es pobre. Una vez absorbido, el TBBFA y/o sus metabolitos parecen distribuirse en la mayor parte de los órganos del cuerpo. En ratas, la semivida máxima de esos compuestos en cualquier tejido fue de menos de dos días y medio.

1.6 *Efectos en mamíferos de laboratorio y en sistemas de prueba* in vitro

La toxicidad oral aguda del TBBFA para los animales de laboratorio es baja. La DL_{50} por vía oral para la rata fue de más de 5 g/kg de peso corporal y la DL_{50} oral para el ratón fue de 10 g/kg de peso corporal. La DL_{50} por vía dérmica para el conejo fue de más de 2 g/kg de peso corporal. Las CL_{50} por inhalación para el ratón, la rata y el cobayo fueron de más de 0,5 mg/litro. Una sola aplicación dérmica de TBBFA en la piel de conejos y cobayos no indujo efectos locales o sistémicos en concentraciones de hasta 3,16 g/kg de peso corporal. El TBBFA no resultó irritante para la piel ni los ojos del conejo. No se observó ninguna reacción de sensibilización en unos pocos estudios realizados en cobayos. El TBBFA también se sometió a ensayos para examinar si tenía actividad cloracnegénica en las orejas del conejo, pero no se observó ninguna reacción de esa naturaleza. En un estudio sobre toxicidad dérmica de tres semanas, en el cual se expuso la piel pelada y raspada de conejos a una cantidad de hasta 2500 mg de TBBFA/kg de peso corporal, sólo se observó un eritema dérmico leve. No se observaron otros cambios relacionados con el compuesto.

Se expusieron ratas a una concentración de hasta 18 mg de TBBFA micronizado por litro (18 000 mg/m^3) durante dos semanas a razón de 4 h/día, 5 días/semana. No se observaron efectos en el peso corporal; tampoco se observaron efectos histopatológicos, ni hematológicos, ni efectos en la química del suero ni en el análisis de la orina.

Dosis orales de hasta 1000 mg de TBBFA/kg de dieta administradas durante 28 días no produjeron efectos adversos. El contenido total de bromo del hígado del grupo de control no resultó diferente del grupo expuesto a dosis altas (1000 mg/kg).

En un estudio sobre toxicidad en ratas se administraron por vía oral durante 90 días dosis de no más de 100 mg de TBBFA/kg de peso corporal; los exámenes del peso corporal, de la composición de la sangre, la química clínica, el análisis de la orina, el peso de los órganos y los exámenes macroscópicos y microscópicos no mostraron ningún efecto adverso.

En un estudio de 90 días realizado en ratones, una dosis oral de 4900 mg/kg de dieta (aproximadamente 700 mg/kg de peso corporal por día) no tuvo ningún efecto adverso; una dosis de 15 600 mg/kg de dieta (aproximadamente 2200 mg/kg de peso corporal por día) provocó disminución del peso corporal, aumento del peso del bazo y reducción de la concentración de eritrocitos, de proteínas del suero y de triglicéridos del suero.

Se realizaron dos estudios sobre teratogenicidad en ratas, uno en el que se administraron con sonda dosis de hasta 10 g/kg de peso corporal desde el día 6 de la gestación hasta el 15 y un segundo en el que se administraron dosis de no más de 2,5 g/kg de peso corporal desde el día 0 hasta el día 19 de la gestación. En el primer estudio, 3 de los 5 animales que habían recibido 10 g/kg murieron, pero no se observaron signos de toxicidad en los animales que habían recibido 3 g/kg. No se observaron efectos teratogénicos. En el segundo estudio no se observó ninguna anomalía.

El TBBFA no tuvo efectos mutagénicos en diversos estudios realizados con cepas TA1535, TA1537, TA1538, TA98 y TA100 de *Salmonella typhimurium* cuyo metabolismo se había activado mediante una mezcla S9 preparada a partir de ratas y hámsters tratados con Aroclor. Las concentraciones ensayadas eran de hasta 10 000 μg/platillo. Los resultados de dos pruebas efectuadas con *Saccharomyces cerevisiae*, con y sin el añadido de una preparación enzimática microsómica tomada de ratas tratadas con Aroclor, también resultaron negativos.

No se comunicaron estudios sobre carcinogenicidad ni toxicidad a largo plazo.

1.7 Efectos en el ser humano

El TBBFA no produjo ninguna irritación dérmica ni sensibilización en 54 personas voluntarias.

No se dispone de estudios ni de otros datos epidemiológicos sobre los efectos en el ser humano.

1.8 Efectos en otros organismos en el laboratorio y en el medio ambiente

El TBBFA no resultó muy tóxico para las algas marinas. En 28 estudios de corto plazo se observaron CE_{50} entre 0,1 y 1,0 mg/litro; las algas de agua dulce no mostraron inhibición del crecimiento, ni siquiera con concentraciones de 9,6 mg/litro.

Se comunicó una CL_{50} aguda a las 48 horas de 0,96 mg/litro para *Daphnia magna*; con 0,32 mg/litro murieron el 5% de los organismos. En un estudio de 21 días, la CE_{50} para la supervivencia y el crecimiento de *Daphnia magna* fue de más de 0,98 mg/litro. Sobre la base de los efectos del TBBFA en la reproducción de dáfnidos en ese estudio de 21 días, se calculó una concentración intoxicante máxima aceptable de 0,3 a 0,98 mg/litro. En misidos (de menos de un día, de 5 y de 10 días de edad), los valores de la CL_{50} a las 96 horas fueron de 0,86, 1,1 y 1,2 mg/litro, respectivamente.

La CE_{50} aguda a las 96 horas (reducción de la deposición de la concha) para ostiones de Oriente se calculó en 0,098 mg/litro, con una concentración sin efectos observados (NOEC) de 0,0062 mg/litro.

Las CL_{50} del TBBFA a las 96 horas para *Lepomis machrochirus* trucha arco iris y *Pimephales promelas* fueron de 0,51, 0,40, y 0,54 mg/litro, respectivamente. Las concentraciones sin efectos para estas tres especies ictiológicas fueron de 0,10, 0,18, y 0,26 mg/litro. Se expuso a *Pimephales promelas* (embriones y larvas) durante 35 días al TBBFA y se observó una concentración intoxicante máxima aceptable (MATC) de 0,16 a 0,31 mg/litro, calculada sobre la base de los efectos adversos en la supervivencia de los embriones y las larvas.

Los niveles sin efectos a los 14 días para el quironómido invertebrado del sedimento *Chironomous tentans* fueron de 0,039, 0,045, y 0,046 mg de TBBFA/litro de agua en sedimentos con contenido bajo, medio y alto, respectivamente, de carbono orgánico.

La mayor parte de los estudios en sistemas acuáticos se han realizado en pH próximos al pKa_2. El comportamiento del TBBFA en aguas ácidas tal vez sea diferente.

2. Conclusiones

2.1 Población general

El TBBFA es muy utilizado y se incorpora en polímeros como reactivo o aditivo pirorretardante. El contacto de la población general con el TBBFA se efectúa por intermedio de productos de esos polímeros y no daría lugar a una ingestión significativa de

TBBFA. Por otra parte, la toxicidad aguda y la de dosis repetidas de TBBFA son muy bajas. La absorción del TBBFA a través del tracto gastrointestinal es mala. El riesgo que para la población general significa la exposición al TBBFA se considera, pues, insignificante.

2.2 Exposición ocupacional

La exposición ocupacional al TBBFA consiste principalmente en exposición a partículas del mismo durante operaciones de envasado o mezcla. El control del polvo mediante la ventilación del local y otros métodos técnicos reducirá el riesgo para los trabajadores. Si el polvo no puede controlarse de manera adecuada, debe utilizarse protección respiratoria.

2.3 El medio ambiente

Las veces que se ha detectado en el medio ambiente, el TBBFA se ha encontrado principalmente en muestras del suelo y de sedimentos. Un factor de bioconcentración relativamente elevado parece quedar compensado por una rápida excreción, de manera que el compuesto no se ha encontrado normalmente en muestras biológicas ambientales.

Los grupos fenólicos del TBBFA se pueden metilar en el medio ambiente y el Me_2-TBBFA resultante es más lipofílico. Este compuesto también se ha hallado en el sedimento y en peces y crustáceos.

2.4 Productos de la descomposición

Se han encontrado DDPB y DFPB como trazas de impurezas en el TBBFA; sin embargo, no se ha demostrado la presencia de congéneres en posiciones 2,3,7,8. En condiciones de pirólisis de laboratorio, se forman DFPB/DDPB a partir del TBBFA.

Un número limitado de estudios han mostrado que durante la elaboración y el reciclado de polímeros que contienen TBBFA como aditivo pirorretardante pueden producirse solamente cantidades muy pequeñas de DFPB/DDPB. Una ventilación apropiada y otros medios técnicos de control pueden prevenir la exposición de los trabajadores.

3. Recomendaciones

3.1 Generalidades

- Los trabajadores de las plantas que fabrican TBBFA y productos que contienen este compuesto deben estar protegidos contra la exposición por medios técnicos de control, y mediante la vigilancia de la exposición ocupacional y la adopción de medidas apropiadas de higiene industrial.
- La exposición ambiental debe reducirse al mínimo mediante el tratamiento apropiado de los efluentes y emisiones en las industrias que utilizan el compuesto o productos del mismo.
- La eliminación de desechos industriales y productos de consumo debe controlarse para reducir al mínimo la contaminación ambiental con este material y con productos de su descomposición.
- Cuando se incinere material tratado con TBBFA, debe hacerse en incineradores de constitución apropiada que funcionen en condiciones óptimas continuas.

3.2 Otros estudios

- La vigilancia de muestras ambientales de TBBFA, Me_2-TBBFA y DFPB/DDPB debe proseguir y, si se encuentran estos compuestos, también debe realizarse una vigilancia en seres humanos.
- Debe medirse la exposición ambiental a partículas de TBBFA que puedan inhalarse durante la respiración; si así lo indicaran los resultados de la vigilancia del medio laboral, debería realizarse un estudio de inhalación a corto plazo en ratas.
- Deberían hacerse estudios sobre la formación de DFPB/DDPB a partir de material tratado con TBBFA durante incineraciones, incendios accidentales y en condiciones de simulación de incendios.
- Deberían realizarse estudios de largo plazo sobre el destino de los polímeros que contengan TBBFA (como resultado de la adición de éste al polímero o bien como resultado de una reacción), especialmente en vertederos.

- Debería estudiarse la conversión ambiental del TBBFA en sus derivados dimetilados, especialmente en sedimentos.

- Deberían proseguir los estudios sobre la reciclabilidad de los polímeros que contengan TBBFA, prestándose atención a los productos de la descomposición.

- Dado que no hay datos, se necesita una prueba *in vitro* adicional con TBBFA para determinar la posibilidad de daños citogenéticos. Si esa prueba resulta positiva, será necesario hacer estudios *in vivo* adicionales. Si las pruebas citogenéticas *in vivo* arrojan resultados positivos, se necesitarán pruebas adicionales de corto y largo plazo.

- Como no hay datos, se necesita una prueba sobre toxicidad reproductiva en ratas.

ETER DIMETILICO DE TETRABROMOBISFENOL A

No hay datos sobre cuya base se pueda hacer una evaluación del éter dimetílico de tetrabromobisfenol A ni respaldar su utilización comercial.

El éter dimetílico de tetrabromobisfenol A no puede evaluarse a menos que se disponga de datos adecuados sobre sus propiedades físicas y químicas, su producción y utilización, su transporte, distribución y transformación en el medio ambiente, sus niveles ambientales y la exposición del ser humano, su cinética y metabolismo en animales y en seres humanos, sus efectos en mamíferos de laboratorio, en seres humanos y en sistemas de prueba *in vitro* y sobre sus efectos en otros organismos en el laboratorio y en el medio ambiente.

ETER DIBROMOPROPILICO DE TETRABROMOBISFENOL A

No hay datos sobre cuya base se pueda hacer una evaluación del éter dibromopropílico de tetrabromobisfenol A ni respaldar su uso comercial.

A partir de los datos disponibles puede concluirse que la toxicidad aguda y de corto plazo del éter dibromopropílico de tetrabromobisfenol A es baja. La sustancia se sometió a pruebas para determinar su mutagenicidad y resultó un mutágeno directo en cepas TA100 y TA1535 de *Salmonella typhimurium*. Sin embargo, los resultados de un ensayo de síntesis no programada de ADN y de una prueba *in vitro* de intercambio de cromátides hermanas resultaron negativos.

Esta sustancia no puede evaluarse a menos que se disponga de datos sobre sus propiedades físicas y químicas, su producción y utilización, su transporte, distribución y transformación en el medio ambiente, sus niveles en el medio ambiente y la exposición humana, su cinética y metabolismo en animales y en seres humanos, sus efectos en mamíferos de laboratorio y en seres humanos y sus efectos en otros organismos en el laboratorio y en el medio ambiente.

BIS(ALIL-ETER) DE TETRABROMOBISFENOL A

No hay una base de datos sobre la cual hacer una evaluación del bis(alil-éter) de tetrabromobisfenol A, ni para respaldar su uso comercial.

A partir de los datos disponibles puede concluirse que la toxicidad oral y dérmica aguda de este compuesto es baja. Los estudios sobre irritación cutánea y ocular en conejos mostraron que la sustancia es levemente irritante para los ojos y la piel.

Esta sustancia no puede evaluarse a menos que se disponga de datos adecuados sobre sus propiedades físicas y químicas, su producción y utilización, su transporte, distribución y transformación en el medio ambiente, sus niveles en el medio ambiente y la exposición humana, su cinética y metabolismo en animales y en seres humanos, sus efectos en mamíferos de laboratorio, en seres humanos y en sistemas de prueba *in vitro* y sus efectos en otros organismos en laboratorio y en el medio ambiente.

BIS(2-HIDROXIETIL-ETER) DE TETRABROMOBISFENOL A

La base de datos es insuficiente para una evaluación del bis(2-hidroxietil-éter) de tetrabromobisfenol A o para respaldar su uso comercial.

A partir de los datos disponibles, hay algunos indicios de que esta sustancia puede hallarse presente en el medio ambiente. Su toxicidad aguda fue baja después de la administración oral y dérmica a ratas y conejos, respectivamente. Su toxicidad aguda por inhalación (1 hora de exposición) en ratas parece ser moderada. Un estudio sobre toxicidad a corto plazo en ratas mostró ausencia de efectos con 1000 mg/kg de dieta, pero se observó un aumento significativo del contenido total de bromo en los órganos. No se observó que la sustancia irritara la piel ni los ojos de conejos. Los resultados de un estudio sobre mutagenicidad en cinco cepas de *Salmonella typhimurium*, con y sin activación metabólica, fueron negativos.

La sustancia no puede evaluarse a menos que se disponga de datos adicionales sobre sus propiedades físicas y químicas, su producción y utilización, su transporte, distribución y transformación en el medio ambiente, sus niveles en el medio ambiente y la exposición del ser humano, sus cinética y metabolismo en animales y en seres humanos, sus efectos en mamíferos de laboratorio, en seres humanos y en sistemas de prueba *in vitro*, y sus efectos en otros organismos en el laboratorio y en el medio ambiente. También se necesita un estudio citogenético *in vitro*.

EPOXI OLIGOMERO BROMADO DE TETRABROMOBISFENOL A

La base de datos es insuficiente para una evaluación del epoxi oligómero bromado de tetrabromobisfenol A y para respaldar su uso comercial.

Se dispone de algunos datos sobre las propiedades físicas y químicas y la producción y utilización del epoxi oligómero bromado de tetrabromobisfenol A, pero dichos datos son insuficientes. Las cantidades de DDPB y DFPB producidas al pirolizar resinas que contienen estos epoxi oligómeros fueron menores que las producidas al pirolizar el TBBFA.

Estas sustancias no pueden evaluarse a menos que se disponga de datos adecuados sobre sus propiedades físicas y químicas, su producción y utilización, su transporte, distribución y transformación en el medio ambiente, sus niveles ambientales y la exposición humana, su cinética y metabolismo en animales de laboratorio y en seres humanos, sus efectos en mamíferos de laboratorio, en seres humanos y en sistemas de prueba *in vitro*, y sus efectos en otros organismos en el laboratorio y en el medio ambiente.

Como la utilización de estos compuestos parece estar aumentando, al menos en el Japón, es esencial que se realicen más estudios.

OLIGOMEROS DE CARBONATO DE TETRABROMOBISFENOL A

No hay ninguna base de datos sobre la cual hacer una evaluación de los oligómeros de carbonato de tetrabromobisfenol A ni para respaldar su utilización comercial.

Los resultados de estudios de mutagenicidad con cinco cepas de *Salmonella typhimurium*, con y sin activación metabólica, fueron negativos para ambas sustancias.

Estas sustancias no pueden evaluarse a menos que se disponga de datos adecuados sobre sus propiedades físicas y químicas, su producción y utilización, su transporte, distribución y transformación en el medio ambiente, sus niveles ambientales y la exposición humana, su cinética y metabolismo en animales y en el ser humano, sus efectos en mamíferos de laboratorio, en seres humanos y en sistemas de prueba *in vitro*, y sus efectos en otros organismos en laboratorio y en el medio ambiente. También se necesitan estudios citogenéticos *in vitro*.

Genetic effects in human populations, guidelines for the study of (No. 46, 1985)
Glyphosate (No. 159, 1994)
Guidance values for human exposure limits (No. 170, 1994)
Heptachlor (No. 38, 1984)
Hexachlorobutadiene (No. 156, 1994)
Alpha- and beta-hexachlorocyclohexanes (No. 123, 1992)
Hexachlorocyclopentadiene (No. 120, 1991)
n-Hexane (No. 122, 1991)
Hydrazine (No. 68, 1987)
Hydrogen sulfide (No. 19, 1981)
Hydroquinone (No. 157, 1994)
Infancy and early childhood, principles for evaluating health risks from chemicals during (No. 59, 1986)
Inorganic lead (No. 165, 1995)
Isobenzan (No. 129, 1991)
Kelevan (No. 66, 1986)
Lasers and optical radiation (No. 23, 1982)
Lead (No. 3, 1977)[a]
Lead - environmental aspects (No. 85, 1989)
Lindane (No. 124, 1991)
Linear alkylbenzene sulfonates and related compounds (No. 169, 1995)
Magnetic fields (No. 69, 1987)
Man-made mineral fibres (No. 77, 1988)
Manganese (No. 17, 1981)
Mercury (No. 1, 1976)[a]
Mercury - environmental aspects (No. 86, 1989)
Mercury, inorganic (No. 118, 1991)
2-Methoxyethanol, 2-ethoxyethanol, and their acetates (No. 115, 1990)
Methyl bromide (No. 166, 1995)
Methylene chloride (No. 32, 1984, 1st edition)
Methyl ethyl ketone (No. 143, 1992)
Methyl isobutyl ketone (No. 117, 1990)
Methylmercury (No. 101, 1990)
Methyl parathion (No. 145, 1992)
Mirex (No. 44, 1984)
Mutagenic and carcinogenic chemicals, guide to short-term tests for detecting (No. 51, 1985)
Mycotoxins (No. 11, 1979)
Mycotoxins, selected: ochratoxins, trichothecenes, ergot (No. 105, 1990)
Nephrotoxicity associated with exposure to chemicals, principles and methods for the assessment of (No. 119, 1991)
Neurotoxicity associated with exposure to chemicals, principles and methods for the assessment of (No. 60, 1986)
Nickel (No. 108, 1991)
Nitrates, nitrites, and *N*-nitroso compounds (No. 5, 1978)[a]
Nitrogen, oxides of (No. 4, 1977)[a]
2-Nitropropane (No. 138, 1992)
Noise (No. 12, 1980)[a]
Organophosphorus insecticides: a general introduction (No. 63, 1986)
Paraquat and diquat (No. 39, 1984)
Pentachlorophenol (No. 71, 1987)
Permethrin (No. 94, 1990)
Pesticide residues in food, principles for the toxicological assessment of (No. 104, 1990)
Petroleum products, selected (No. 20, 1982)
Phenol (No. 161, 1994)
d-Phenothrin (No. 96, 1990)
Phosphine and selected metal phosphides (No. 73, 1988)
Photochemical oxidants (No. 7, 1978)
Platinum (No. 125, 1991)
Polybrominated biphenyls (No. 152, 1994)
Polychlorinated biphenyls and terphenyls (No. 2, 1976, 1st edition)[a] (No. 140, 1992, 2nd edition)
Polychlorinated dibenzo-*p*-dioxins and dibenzofurans (No. 88, 1989)
Progeny, principles for evaluating health risks associated with exposure to chemicals during pregnancy (No. 30, 1984)
1-Propanol (No. 102, 1990)
2-Propanol (No. 103, 1990)
Propachlor (No. 147, 1993)
Propylene oxide (No. 56, 1985)
Pyrrolizidine alkaloids (No. 80, 1988)
Quintozene (No. 41, 1984)
Quality management for chemical safety testing (No. 141, 1992)
Radiofrequency and microwaves (No. 16, 1981)
Radionuclides, selected (No. 25, 1983)
Resmethrins (No. 92, 1989)
Selected synthetic organic fibres (No. 151, 1993)

---

[a] Out of print